COSMOLOGY
WITH MATLAB

Revised with MATLAB Live Scripts

Second Edition

Other World Scientific Titles by the Author

Lectures in Particle Physics
ISBN: 978-981-02-1682-5
ISBN: 978-981-02-1683-2 (pbk)

Physics at Fermilab in the 1990's: Proceedings of the Workshop
ISBN: 978-981-02-0103-6

Beams and Accelerators with MATLAB: With Companion Media Pack
ISBN: 978-981-323-746

Stars and Space with MATLAB Apps: With Companion Media Pack
ISBN: 978-981-12-1602-2
ISBN: 978-981-12-1635-0 (pbk)

The Physics of Experiment Instrumentation Using MATLAB Apps:
With Companion Media Pack
ISBN: 978-981-12-3243-5
ISBN: 978-981-12-3383-8 (pbk)

One Hundred Physics Visualizations Using MATLAB
Second Edition
ISBN: 978-981-12-9561-4

COSMOLOGY WITH MATLAB

Revised with **MATLAB** Live Scripts

Second Edition

Dan Green
Fermi National Accelerator Laboratory, USA

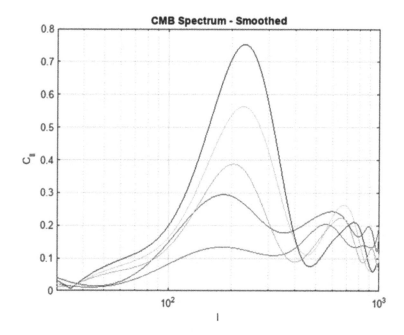

 **World Scientific**

NEW JERSEY · LONDON · SINGAPORE · GENEVA · BEIJING · SHANGHAI · TAIPEI · CHENNAI

Published by

World Scientific Publishing Co. Pte. Ltd.

5 Toh Tuck Link, Singapore 596224

USA office: 27 Warren Street, Suite 401-402, Hackensack, NJ 07601

UK office: 57 Shelton Street, Covent Garden, London WC2H 9HE

Library of Congress Control Number: 2024945453

British Library Cataloguing-in-Publication Data
A catalogue record for this book is available from the British Library.

COSMOLOGY WITH MATLAB
Revised with MATLAB Live Scripts
Second Edition

ISBN 978-981-98-0103-9 (hardcover)
ISBN 978-981-98-0152-7 (paperback)
ISBN 978-981-98-0104-6 (ebook for institutions)
ISBN 978-981-98-0105-3 (ebook for individuals)

For any available supplementary material, please visit
https://www.worldscientific.com/worldscibooks/10.1142/14047#t=suppl

Desk Editor: Carmen Teo Bin Jie

Typeset by Stallion Press
Email: enquiries@stallionpress.com

Preface

"The Cosmos is all that is or was or ever will be. Our feeblest
contemplations of the Cosmos stir us — there is a tingling in the
spine, a catch in the voice, a faint sensation, as if a distant
memory, of falling from a height. We know we are approaching
the greatest of mysteries."

— Carl Sagan

"The more of us that feel the Universe, the better off we will be
in this world."

— Neil deGrasse Tyson

The text is now eight years old, and much has transpired during that time.
The MATLAB tools have evolved from scripts, to Apps and, at present, to
Live code. The Live package is preferred because it combines text, figures
and equations with MATLAB code all in a single package. The numerical
results of that code, formerly shown separately, also appear in line and
in this way the user can vary the parameters of the specific problem and
explore immediately how the solutions vary in response. For this reason
the Live schema is used exclusively in this edition. In the previous edition
the MATLAB script appeared in an Appendix. With Live script all output
appears in the same place and is all updated immediately if any parameter
is changed by the user.

The physics landscape for cosmology has also evolved significantly. The
Nobel prize in 2006 rewarded the discovery of small perturbations in temperature, at the parts per million level of the extreme isotropy of the Cosmic
Microwave Background (CMB). The basic isotropy is now thought to indicate a period of rapid expansion of the Universe, called "inflation".

The 2011 Nobel prize was for the observation, using supernovae as "standard candles", of the accelerating expansion of the universe. This expansion is now ascribed to the existence of "dark energy".

In 2013 the prize was awarded for the discovery of the Higgs boson, a fundamental scalar, and the first and only such fundamental particle. Its discovery completes the Standard Model of Particle Physics, SMPP. Indeed, the Higgs boson has the quantum numbers of the vacuum, as would the field for "inflation".

In 2017 the prize went for the observation of gravitational radiation using gravity wave detectors. In 2019 the prize was awarded for the explication of the structures in the CMB and the subsequent emergence of the cosmic "standard model" where the Universe is composed of matter, photons, dark matter and dark energy.

In 2020, the prize was given for the exposition of the nature of the singularities of General Relativity (GR), black holes.

These awards have altered the choices made in the topics that are addressed in this revised volume.

Cosmology has recently also made great strides in the accuracy of the data which have become available. A "Standard Model of Cosmology", SMC, has been developed which successfully describes the Universe in terms of only a handful fundamental parameters. However, with more accurate data a tension in the results of extracting the cosmological parameters using different methods has arisen which may indicate that the present SMC is incomplete or oversimplified.

The text is divided into 9 sections. The first introduces MATLAB. In section 2 historical topics in weak field General Relativity (GR) and other relevant physics topics are discussed. Section 3 explores the simplest cosmic metric, the Robertson-Walker (RW) metric which is defined by a single scale factor, a(t). In Section 4 the history of the Universe from the hot Big Bang (HBB) is sketched out up to the CMB. Section 5 deals with the later evolution in the HBB epochs.

In Section 6 inflation is introduced, first with a constant Hubble parameter, H, and then with a more realistic scalar field realization of the inflationary paradigm. Inflation explicitly solves the issues with HBB cosmology and fundamentally changes the evolution of horizons from the HBB model to the present SMC. Not only does inflation solve the problems of HBB cosmology, but it predicts quantum fluctuations at early times as discussed in Section 7. The reader has by then developed the tools needed to evaluate the compatibility of any particular model with the existing experimental constraints as they evolve and improve with time.

The fluctuations evolve and lead to the prediction of classical scalar and gravity wave perturbations. The small scalar fluctuations have, indeed, been seen imprinted on the CMB and the evolution of large scale structure in the matter distributions which are displayed in Section 7 for the CMB and Section 8 for the Large Scale Structure (LSS) of matter. Section 9 ends the text with a second look at the Higgs as the cause of inflation, not because it is a favored model but because it is the simplest way to try to connect the energy scales of the SMPP with the SMC, and is a minimalist approach from that point of view.

In Appendix A, some brief descriptions of some of the very extensive suite of MATLAB tools are provided. Then in Appendix B a table of formulae applicable if a single GR source dominates, matter, radiation or dark energy. A Table of the symbols used in the text is given in Appendix C. References are not provided, and the reader is encouraged to use the many available internet search tools to explore a specific topic. A single specific suggestion is given in Appendix D. Appendix E collects the extensive number of acronyms used in the text.

Contents

Chapter 1

Introduction to MATLAB Tools

"Mathematics expresses values that reflect the cosmos, including orderliness, balance, harmony, logic, and abstract beauty."
— Deepak Chopra

"Pure mathematics is, in its way, the poetry of logical ideas"
— Albert Einstein

"I don't think we have reached a point where art really translates into science. Perhaps for some people, having good visuals can help translate into science."
— Lisa Randall

1.1 Startup

The "Live" version of MATLAB is used in the text because all elements — the code, the computation, the variables to be studied, and the plots and numerical or symbolic output — are all available coherently. This text can be read statically, but using the included MATLAB scripts will give the user the full Live experience, which is the intent of this volume. The user can run each of the Live scripts and then play with the parameters of the specific problem. This is accomplished by changing "edit fields", "numeric sliders", or "dropdown" menus. Any change causes a re-execution of the script.

First, install and open MATLAB. Opening MATLAB brings up the Command Window, shown in Fig. 1.1. The prompt for input is >>.

Other windows, such as the Workspace window that shows all active variables and the Command History that shows all past session commands, are docked in this view with the view chosen using the Layout menu. The Editor opens with the "New" or "Open" tab for new or existing code.

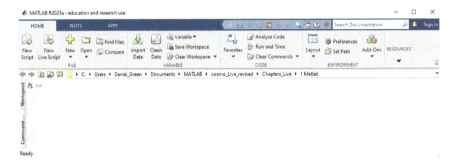

Figure 1.1:   MATLAB Command Window.

The Search Documentation provides a complete search of the extensive MATLAB documentation and often gives illustrative examples. For example, a search for documentation yields Fig. 1.2.

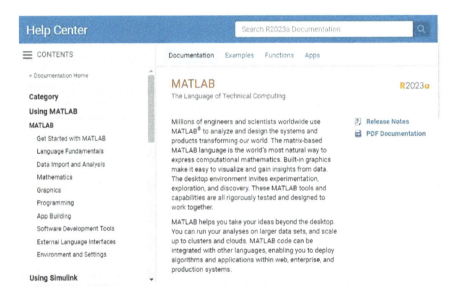

Figure 1.2:   Tutorial on using MATLAB from the Search Documentation window.

Many tools are included with MATLAB. This fact drives the decision in this text to use MATLAB tools to avoid unnecessary algebraic tedium. Most problems are solved explicitly using symbolic math tools. In general, numerical results use the extensive MATLAB library of special functions and of the numeric solvers of ordinary and differential equations. Plots of the results are displayed. The types of plots and examples are invoked with the "Plots" tab on the home page. The user of a Live script is typically given a choice of parameters to set, and the results of any changes are immediately evaluated and plotted. The text is designed to focus on concepts and not solutions or calculations, leaving those to MATLAB utilities.

1.2 Onramp Course

To refresh/get started, the MATLAB "Onramp" is also very useful, with an outline of the topics covered given in Fig. 1.3. Login to Mathworks and then search for "Onramp" for a good MATLAB introduction. The "resources" tab in the Command window also gives access to useful tutorials.

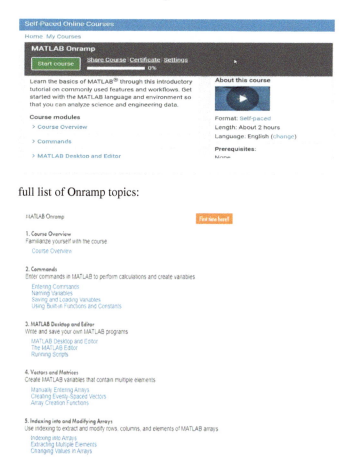

Figure 1.3: Screenshot of the "Onramp" MATLAB introduction.

1.3 Symbolic Math Tools: Introduction

Then, one can run the Live script "Symbolic_X_LV", which is appended, for an introduction to MATLAB symbolic logic — tools which are heavily used in the text. The aim here is to use the symbolic math tools to avoid much of the algebraic equation work and rather concentrate first on the physics and then the parameter dependence of the solutions. (see Figure 1.4).

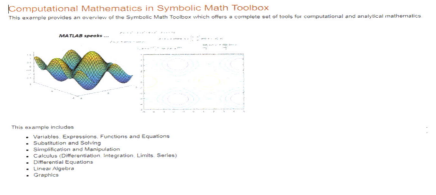

Figure 1.4: Screenshot of the initial "Symbolic_X_LV" output.

It is easiest to separately run the live script "Symbolic_X_LV" in order to see an introduction to MATLAB symbolic logic, such as syms (defining symbolic variables), solve (solve algebraic equations), subs (substitute), simplify (simplify a symbolic expression), factor (factorize), diff (differentiate), int (integrate), and dsolve (solve a differential equation). These and other symbolic tools are used heavily in the text so that a quick introduction is a good first step.

In the following example, the user can type in any function of x to see the derivative of that function:

```
% Symbolic_X_LV
syms f(x)
f(x) = str2sym('cos(x)') ;
```

```
f(x) = cos(x)
```

```
diff(f(x))
```

```
ans = -sin(x)
```

1.4 Script Format: Editor

The opened scripts are displayed in the Live Editor. Apps have a distinct Editor. The formatting used by the editors is very useful for understanding the flow of computation. Data can be input to the Command Window by .keyboard with prompt >>. Text strings are in purple. Comments are, %, in green. Symbolic variables and text characters are in purple, and operations are executed in black. Flow control is in blue. Printing to the Command Line also appears inline for "Live" scripts which allows one to view the output in the same location as the script. This feature of "Live" scripts is very useful in following the flow of the calculation and it will typically be used in this text. Live Editor comments on warnings and errors appear on the right. If a statement is not followed by a %, the printout is inline with the Live script. The Live Editor "INSERT" tab allows added Control (Edit Fields, DropDown menus, and Numeric Sliders) to be inserted which enables user input. Text and Equations can also be added to the Live code.

```matlab
global xx ;    % global (common) variables are blue (unfavored now in
  Matlab, use function arguments to pass variables)
for i = 1:2     % controls are in blue - indented by loops
    yy(i) = i *sin(i);
end
% any syntax errors appear in red, with script explanation on hover
% comments are in green
xx = linspace(0,2 *pi,100); % executables are black
```

A short answer to a query about a MATLAB function can be found using the "help" query in the Command Window or Live script. For example, help on the sin function. For more extensive documentation, use the Search Documentation window in the Home tab. MATLAB has an extensive suite of special functions made available. Almost any mathematical function used in physics can be found in MATLAB and used as a library.

```matlab
help('sin')
```

sin Sine of argument in radians.

sin(X) is the sine of the elements of X.

See also asin, sind, sinpi.

Documentation for sin

Other uses of sin

1.5 Other Languages

MATLAB is available at many colleges and universities via a site-wide license. There is also a reduced- cost student version. In addition, there are ways to make other languages compatible with MATLAB. Python is a specific example. A "Cheatsheet" for MATLAB/PYTHON is supplied in Appendix A.

1.6 Cheatsheets

There are "cheatsheets" for a lookup of MATLAB utilities. A cheatsheet for general MATLAB operation is presented in Appendix A.

Chapter 2

Physics Which is "Recently" Understood

"Because there is a law such as gravity, the universe can and will create itself from nothing".

— Stephen Hawking

"The black holes of nature are the most perfect macroscopic objects there are in the universe. The only elements in their construction are our concepts of space and time".

— Subrahmanyan Chandrasekhar

The MATLAB "Live" system combines script and output, which is its great strength. The scripts are a companion pack with the text. However, there are many lines of script. Due to a desire to limit the length of the text, the scripts are truncated to include only the salient points, such as "explanatory comments" and when new MATLAB tools are first used. In this way, the script for plots will often be omitted as being repetitive.

2.1 Tests of Weak Field GR

There were arguably three major advances in physics in the twentieth century: quantum mechanics, SR + GR, and cosmology. Gravity is central to cosmology so the evidence for GR as the theory of gravity is explored first. In GR there are few closed-form solutions and superposition is impossible since gravitational energy gravitates. This means one cannot build up more complex solutions by adding point solutions, as is done in electromagnetism. However, the basic idea is familiar. There is a source term comparable to charge. In this case, the sources are mass and pressure, ρ and p/c^2, which are both dimensionally energy densities divided by c^2 and are elements of the stress-energy tensor, T_{uv}. If a quantum theory of gravity were to emerge in the future, the quanta of the field would have spin = 2, similar to the spin 1 quanta, photons, of the vector electromagnetic field. The classical field equations are differential equations connecting the source tensor to the space-time metric, g_{uv}. The tensor $R_{\mu\nu}$ is a second-order differential of the metric. Unfortunately, they are non-linear equations and only a few closed-form solutions are known:

$$R_{uv}(g_{uv}) + (\Lambda - R/2)g_{uv} = (8\pi G/c^4)T_{uv} \tag{2.1}$$

The R_{uv} is a second-rank tensor acting on the metric with differential operators. The term with the parameter Λ, proportional to the metric itself, is an interesting one. It was a term added by Einstein to be "attractive" and insure a static Universe but then retracted as what he called his biggest mistake. However, it now has a new life as a "cosmological constant" and can describe the repulsive "dark energy" (DE), which appears to be the majority of the energy in the present Universe. It possesses the quantum numbers of the vacuum and is proportional to the space itself through the metric. It can be either repulsive or attractive, although for cosmologists it is taken to be repulsive. The field equations say that the geometry, the left-hand side of Eq. (2.1), is caused by the energy content of the space, the right-hand side. Given a metric which solves some problem, the geodesic equations for the path of a test particle can be determined.

Electromagnetism also has a field tensor, $F^{\mu\nu}$, containing the electric, E, and magnetic, B, fields. The contracted field tensor is proportional to the field sources, the electromagnetic current, J^ν. The force carriers of the standard model of particle physics (SMPP) are all vector particles, like the photon:

$$\partial_\mu F^{\mu\nu}(E, B) = J^\nu \tag{2.2}$$

The weak field limits of the field equations should approach the Newtonian version of gravity and lead to the familiar Poisson equation relating the gravitational potential ϕ to the matter density ρ. Note that pressure is a source in GR. The dimensions of p are energy density, as are the dimensions of ρc^2. The stress–energy tensor and the geodesic equations in weak GR are

$$T = \begin{bmatrix} \rho & 0 \\ 0 & \vec{p}/c^2 \end{bmatrix}, \quad \nabla^2\phi = 4\pi G\rho \tag{2.3}$$

The addition of DE, or a cosmological term, Λ, means that the Poisson equation becomes $\nabla^2\phi = 4\pi\rho/M_p^2 - \Lambda$, where M_p is the Planck mass, to be defined later.

The earliest experimental tests of GR can all be understood by fairly basic arguments. Three early tests of GR were all small effects and had to do with the red shift of photons climbing out of a gravity well (Earth), the deflection of light travelling near the Sun during a solar eclipse, and the perihelion advance of Mercury. In the case of the perihelion advance, the GR geodesic equations added a small force which was not inverse square which therefore made the eccentric orbit no longer re-entrant. The red shift follows from energy conservation. The photon energy at some height is $\varepsilon = \hbar\omega$. After rising by a height Δy from near the Earth's radius, R_E, the photons lose energy and their frequency is shifted toward the red. The Schwarzschild radius is defined to be $R_s = 2GM/c^2$. The red shift establishes that light, having energy, has gravitational mass. A very schematic view of the effect is shown in Fig 2.1.

$$\Delta\omega/\omega = -(\Delta y/R_E)(R_s/2R_E) \tag{2.4}$$

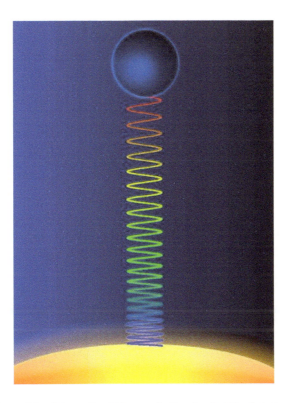

Figure 2.1: Schematic of the gravitational red shift of photons.

Another early test of GR was the observation that light was deflected by passing near the Sun. As with the red shift, light, having mass, was "attracted" to the Sun. For now, one can simply accept the resulting equation of motion for photons, where $u = 1/r$:

$$\frac{d^2}{d\theta^2}u + u = 3\mathrm{GMu}^2 : \quad (u \sim \sin(\theta)/r, \quad 3\mathrm{GMu}^2 \sim 3\mathrm{GM}(\sin\theta/R)^2 \quad (2.5)$$

An approximate solution comes from taking the small term due to GR and using a Newtonian straight line approximation for u. The experimental result for the Sun at eclipse is a deflection of $1.75''$. In general, $\theta \sim 2R_s/R$, which is small because R_s is a few kilometers, much smaller than the solar radius, R. The user-chooses the ratio. Orbits are shown in Fig. 2.2. The blue line on the plot is an undeflected Newtonian photon. The colored lines are the approximate GR light deflection for the user chosen value of R/R_s.

By the way, it is important to the purpose of this text that the user sees the changes in the results as a function of the change in parameters. Due to this, the "close all" command is commented out. Any change in the user input, MR in this case, immediately overlays a new plot, for comparison. The plot can be cleared at any time by rerunning the script.

```
% GR light deflection
%close all;
clear all;
% light deflection at min radius R of a body of mass M
MR = 0.01; % Rs/R, point of closest approach
% Newtonian straight line u = 1/r = sin(phi)/R
% GR not in closed form, start with u = 1/r = sin(phi)/R
% Newtonian asymptotes at phi ~ 0, no deflection
% GR asymptotes, small angle approx, at +- Rs/R, total deflect 2Rs/R
R = 1;    % units of solar radius
sp = linspace(0.1,1);
cp = sqrt(1.0 - sp .*sp);
% Newtonian straight line
rnew = R ./sp;
xnew = rnew .*cp;
ynew = rnew .*sp;
% approx GR with curvataure and asymptotes
uu = sp ./R + (MR/R^2);
rGR = 1.0 ./uu;
xG = rGR .*cp;
yG = rGR .*sp;
plot(xnew,ynew,-xnew,ynew,'-b')
hold on
```

```
plot(xG,yG,-xG,yG)
title('Small Angle Orbits for Light, Newtonian and GR')
xlabel('x/R')
ylabel('y/R')
```

```
axis([-10 10 0  .6 1.1])
```

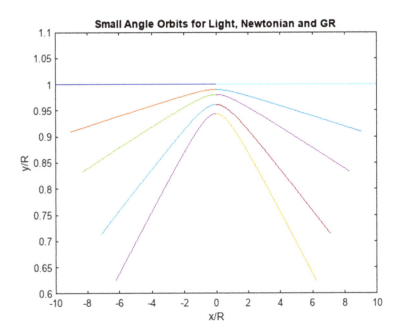

Figure 2.2: Schematic view of an orbital trajectory for light in GR.

The material particle GR geodesic equation of motion is as follows:

$$\frac{\mathrm{d}^2}{\mathrm{d}\theta^2}u + u = R_s/2h^2 + 3R_su^2/2, \quad u = 1/r \tag{2.6}$$

The constant of the motion is now h which is closely related to the Newtonian angular momentum L: $h = (L/c)/(1 - R_s/r) \sim L/c$, $L = r^2(\mathrm{d}\theta/\mathrm{d}t)$. Note the presence of an additional term with respect to the Newtonian case. This means the orbits are no longer defined by an inverse square force law and the orbits will not be re-entrant. This GR term is the origin of the perihelion advance which is the third initial historical test of GR, after the red shift of light and the deflection of light by the Sun.

The GR equation for light occurs in the special case where $h \to \infty$ which sets the classical centrifugal force term to 0.

For a planet with eccentricity e and semi-major orbital axis a, there is an advance of the perihelion of $3\pi(R_s/a)/(1 - e^2)$ which for Mercury is $0.1''$ per rotation or $43''$ per century. This prediction of GR cleared up a long-standing problem in astronomy and was a convincing proof of the GR hypothesis in the weak field limit. The early tests were all measuring small quantities due to factors of $\sim R/R_s$.

A very recent test, reported in 2023 and shown in Fig. 2.3, was made at CERN on the gravitational response of atoms of positrons bound to antiprotons (anti-hydrogen) found that such atoms fell in the Earth's field just as do normal atoms. This is another weak field GR test; all matter has energy and gravitates universally.

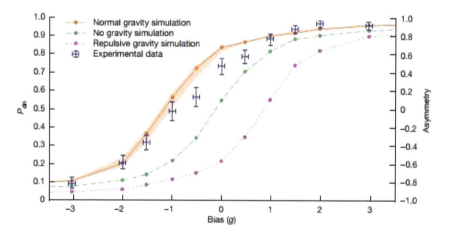

Figure 2.3: Data from a CERN experiment on anti-hydrogen in a gravity field.

2.2 Successful Solar Modeling

Until the twentieth century, the Sun was a problem because the solar energy released could not be explained. The geologists and evolutionary biologists felt that the Earth must have existed for a very long time. The physicists did not see how that was possible. It was only with the advent of nuclear physics that the fusion processes which drive the Sun were understood and a lifetime of the Sun of billions of years became understandable.

There are many models of the Sun. A simple approach is adopted here and the results are compared to a reasonably successful model which fits solar data. The aim is to go beyond the uniform density assumption. To begin, the Sun is treated as an ideal gas with the majority constituents of its mass being protons and helium ions. The hydrogen and helium are ionized at core temperatures corresponding to MeV energies compared to the eV ionization energies. The core is then a neutral plasma of non-relativistic, NR, protons, electrons, and helium nuclei. The energy density is the product of the number density, n, times the mean thermal energy, $(3kT)/2$, the Boltzmann result for the mean. Pressure arises due to the plasma temperature and the photon radiation pressure:

$$PV = NkT, \quad n = N/V$$

$$m\langle v^2 \rangle / 2 = 3kT/2 \tag{2.7}$$

$$P = \rho kT / (\mu m_p) + 4\sigma T^4 / 3c$$

The radiation pressure follows from the Planck distribution for energy. The Stefan–Boltzmann constant is $\sigma = 5.67 \times 10^{-8} \text{W/m}^2\text{K}^\circ$. The phase space for photons yields an energy density, u, factor x^3 times the Bose–Einstein weight factor. The integrated radiation energy density scales as T^4:

$$du(T)/dE = (kT)^3 / \left[\pi^2 (\hbar c)^3 \right] \left\{ x^3 / (e^x - 1) \right\}, \quad x = E/kT$$

$$u = 4(\sigma T^4)/c = (kT)^4 \pi^2 / [15(\hbar c)^3] \tag{2.8}$$

The mass density is related to the number density by μm_p, where μ is the mean molecular weight. The mean molecular weight is the average mass of a particle in the gas so that $\mu m_p \sim 0.62$ for the mix of protons and helium nuclei of a young star when fully ionized, $1/\mu \sim 2X + 3Y/4$, where X is the proton mass fraction and Y is the helium mass fraction. The mean mass is less than a proton mass because of the existence of electrons in the ionized mixture. The neutral fraction when the gas is not ionized is $1/\mu \sim X + Y/4 = 1/1.3$ larger than the proton mass due to the heavier helium with A of four.

The simplest solar model assumes a steady state stable solution for the star which is characterized by a density ρ, a pressure, P, and a temperature, T, all functions of the radius r from the solar center. The star is in equilibrium between the gravitational pressure and the outward pressure due to the gas and the radiation. In the conventional notation used here, $M(r)$, $\rho(r)$, $P(r)$, and $T(r)$ denote the mass within r and the density, pressure, and temperature at r. A simplification is to assume that the pressure can be dominated by the contribution due to either the gas or the radiation alone.

The core where fusion occurs is only about 10% of the mass of the Sun. In fusing protons into helium, only about 0.007 of the rest mass is converted to energy. Typical binding energy is 7 MeV, while typical light nuclear mass is 1000 MeV, ratio = 0.007. An estimate of the total possible energy release is then $0.1 \times 0.007 \times M_o c^2$ or about 1.3×10^{44} J. Comparing to the solar luminosity, 3.8×10^{26} W, a crude estimate of the solar lifetime is then about 10 billion years. The Sun is expected to burn stably for a good fraction of the time the Universe has been in existence, currently estimated to be about 13.8 billion years. This simple order of magnitude estimate allows for a long lifetime for the Sun once the fusion process is understood. The rest of the model is simple classical physics.

The basic equations for the simple solar model and the boundary conditions are, ignoring the radiation pressure here in the interest of simplicity

$$dM(r)/dr = 4\pi r^2 \rho(r) \qquad\qquad M(0) = 0, \quad M(R) = M_T$$
$$dP(r)/dr = -GM(r)\rho(r)/r^2 \quad P(R) = T(R) = \rho(R) = 0 \qquad (2.9)$$
$$P(r) = \rho(r)\mathrm{kT}(r)/\mu m_p$$

The radiation pressure term could be substituted for the temperature term for the pressure, to get a feeling for the relative contribution. The equation for the mass r dependence on density is clear, while the dependence of pressure on G, M, and density was mentioned previously, $dP/dr \sim G\rho^2 r$, $\rho \sim M/r^3$. The temperature dependence follows from the ideal gas law for pure protons and is altered for a mixture of ions. The boundary conditions at the core and the surface define the meaning here of solar radius, R. The quantity μ is the mean molecular weight, $\mu = \langle m \rangle / m_p$.

The specific heat is defined to be the amount of heat needed to raise the temperature of an object one degree. If the processes are adiabatic, then there is no net heat flow. Rising gas expands and cools. Falling gas contracts and heats. An adiabatic process has PV^γ constant. The factor γ is the ratio of dP/P divided by dV/V which is the ratio of the specific heats and has a value of $5/3$ for an ideal gas. This is the case for simple convective dominance.

In the case of radiative dominance, the photons scatter off the charged protons, electrons, and H ions. The scattering is characterized by an opacity, κ, and a luminosity, $L(r)$. For a density ρ and path length z, the photon intensity falls exponentially with z with an exponent of $-\kappa\rho z$. The mean free path is $n\sigma = \kappa\rho$, where σ is the cross-section for light scattering off the solar medium. The convective and radiative temperature contributions are:

$$\mathrm{d}T(r)/\mathrm{d}r = \{(\gamma - 1)/\gamma\}\mu m_p GM(r)/r^2$$
$$\mathrm{d}T(r)/\mathrm{d}r = -3\langle\kappa\rangle\rho(L(r)/4\pi r^2)/16\sigma T^3 \tag{2.10}$$

The luminosity behavior, $\mathrm{d}L(r)/\mathrm{d}r$, follows that of the mass but modified by an energy production factor, ε, which, since it represents fusion, is highly temperature dependent and thus biased toward the inner, high temperature, core. The opacity, with dimension m^2/kg, is set to a constant times ρ divided by $T^{-3.5}$. The fusion energy production factor is taken to be that for p-p reactions. The user can change the factor, ε, as desired. For the Sun, the luminosity- to- mass ratio implies a mean power production of about $0.2\,\mathrm{mW/kg}$. At the core, the power production is the highest, about 4.3 MW/kg, due to a core number density of protons of about $9 \times 10^{31}/\mathrm{m}^3$:

$$\mathrm{d}L(r)/\mathrm{d}r = 4\pi r^2 \rho(r)\varepsilon = [\mathrm{d}M(r)/\mathrm{d}r]\varepsilon \tag{2.11}$$

Global solar properties are shown in Table 2.1. The differential equations are solved as a boundary value problem (BVP) using the MATLAB utility "bvp4c" which uses the boundary values to solve for the interior parameters in the purely radiative case. The purely convective case is a poorer fit to the "data". Applying the boundary values defines the core values of pressure and temperature. For the Sun, the fusion occurs in about the first inner ~30% of the radius, where radiation pressure dominates. Indeed, the convective region for the Sun is only the outermost radii, near the solar surface. The other method of assuming a core pressure and temperature and integrating to the solar surface is not very robust since the problem

Table 2.1: Solar properties.

Property	Symbol	Value
Radius	R(m)	6.9×10^8
Mass	M (kg)	2.0×10^{30}
Luminosity	L (W)	3.8×10^{26}
Surface temperature	T (°K)	5770
Solar constant (Earth)	f (W/m^2)	1.37×10^3
Schwarzschild radius	R_s (km)	3.0
Number density − mean	n (1/m^3)	1.2×10^{23}
Mass density − mean	ρ (kg/m^3)	1.4×10^3
Pressure = GMρ/R − mean	P (Pa = Nt/m^2)	3×10^{14}
Core temperature = T_c	Tc (°K)	1.5×10^7
Core pressure = P_c	Pc (Pa)	2.6×10^{16}
Core mass density = ρ_c	ρ_c (kg/m^3)	1.5×10^5
Hydrogen and helium mass fraction,	X, Y	0.7, 0.28
X and Y $X + Y \sim 1$		
X and Y, core values, ionized	X, Y	0.35, 0.65
Mean molecular mass fraction, ionized	μ	1.30 H, 0.62

is not really an initial value problem. The result is a much improved representation of the Sun compared to trying initial value solutions, although this particular formulation, about as simple as can be and still able to be a reasonable match to the "data", is slightly unstable.

Scaled variables are used to reduce the variations. They are m, x, p, and t defined as $M = mM_T$, $r = xR$, $P = p(GM_T^2/4\pi R^4)$, $T = t(\mu m_p/k)(GM_T/R)$. The equations are solved for m, p, t, l, while ρ is derived using the ideal gas law. This BVP approach is much slower and less robust than the more typical initial value problem solver, "ode45". However, it arrives at a better representation of the solar "data".

Given the agreement with the "data", we can assume that the physics of suns is fairly well understood, and one need not rely only on the simplest uniform density solar model. A uniform density stellar model is quite far from reality, and the core is quite evident in the plots given Fig. 2.4.

```
% treat sun as a BVP -  opacity included - sun
% the set of 4 scaled eqs for m,p,l and t are used and rho derived
from ideal gas
global G M R Pscale Tscale rhoscale Lscale beta ikey
```

```
% constants
c = 3e8; % light veocity
G = 6.67e-11 ; % newton constant
mp = 1.67e-27 ; % proton mass kg
sig = 5.67e-8 ; % S-B in W/(m^2deg^4), sig/c is an energy density/T
^4
kb = 1.38e-23 ; % Boltz const J/deg
Ro = 6.96e8 ; % sun radius m
Mo = 2e30 ; % sun mass kgm
Po = 1.29e16 ; %sun central pressure nt/m^2
To = 1.57e7 ; % sun central temp deg K
Lo = 3.8e26 ; % luminosity in W
Teff = 5770 ;  % surface temp , Lo =4*pi*Ro^2*sigma*Teff^4
rhoo = 1.58e5  ; % central sun density kg/m^3
% Solar model from Schwarzchild - assumed to be a good model
% star defined by total mass and radius and bc - need for lumi
R = Ro ; % specialize to the sun
M = Mo;
% dimensionfull scales for dimensionless param m, p, t, l
Pscale = (G .*M .^2) ./(R .^4 .*4 .*pi);
mu = 0.62; % hydrogen by weight, H + He - ionized
Tscale = (mu .*mp .*G .*M) ./(kb .*R);
rhoscale = (mu .*Pscale .*mp) ./(kb .*Tscale); % derived from ideal
gas
% rho = rhoscale(p/t)
Lscale = Lo ; %(G .*M .^2 .*c) ./R .^2;
beta = -(3.0 .*Lscale .*Pscale .*mu .*mp) ./(64 .*pi .*Tscale .^5 .
*R .*kb .*sig);
% coefficient in dt/dx = beta*(p*l)/(x^2t^4)*beta in radiative needs
 kappa
xx = linspace(0,1,20); % linear initial guess, satisfies bc
solinit = bvpinit(xx,solar_init);
% evaluate the scaled solution
Pc = Pscale .*real(sol.y(2,1));
Pc = Pc./Po; % scale to the sun
Tc = Tscale .*real(sol.y(3,1));
Tc  = Tc./To ;
mu = 0.62; % hydrogen by weight, H + He
eps = 2.46e6 .*rhsc .*0.36 .*T6 .^-0.666 .*exp(-33.8 .*T6 .^-0.333);
% eps in erg/gm*sec -> Watt/kg
eps = eps .*1e-4;
```

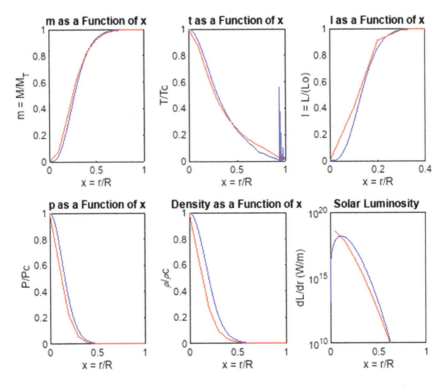

Figure 2.4: Plots of the simple solar BVP solution, blue, and pseudo data from a successful and detailed solar model, red.

Note that this analysis assumes the star has formed and is in a state of equilibrium. That state may last for billions of years. This does not answer the question of how these stars formed in the early Universe. Indeed, new data from new devices inform on the speed of the formation of large-scale stellar structures in the early Universe as is explored in later sections.

2.3   Solar Collapse Prediction

Stars are composed of mostly hydrogen undergoing fusion reactions which release enormous amounts of energy. Protons and neutrons fuse into deuterium and finally end up as helium nuclei and release a large binding energy. For example, $H^2 + H^3 \rightarrow \text{He} + n + 17.6$ MeV, with a large increase in kinetic energy. The process can occur rapidly because the temperature is high enough to overcome the Coulomb repulsion of the protons. The radiation pressure of the photons is balanced by gravitational attraction, and stars can be stable for billions of years. However, ultimately gravity wins.

As stars run out of fusion fuel, the core will begin to contract under their self-gravitational interactions because the radiation pressure can no longer balance the gravitational attraction. The fate of the star depends on quantum mechanics, specifically the Fermi pressure which is the result of the Fermi exclusion principle. At issue is whether the contraction and ultimate collapse can be evaded. The gravitational binding energy is $U = (3/5)(GM^2/Rc^2)$ which diverges at $R = 0$ unless opposed by other mechanisms.

For a uniform density, ρ, star the of mass M and radius R, the gravitational binding energy U is

$$U = -(4\pi)^2 G\rho^2 R^5 \sim -GM^2/R \tag{2.12}$$

This energy is dimensionally $-GM^2/R$. The pressure, P, is due to compression. There is a change in gravitational self-energy as a function of volume, V, which is the pressure, P. It scales as the inverse fourth power of the radius. Expressed as a function of the total number of nucleons N, it scales as N squared:

$$P_G = \frac{\partial}{\partial V}U \sim GM^2/R^4 \sim N^2/V^{4/3} \tag{2.13}$$

With all fusion fuel gone, there is nothing to oppose this pressure and avert a collapse of the star except the Fermi pressure of the fermions, the electrons and nucleons, in the star. The Fermi wave vector is $K_F = (3\pi^2 n)^{1/3}$, where n is the number density of fermions. The Fermi energy itself, E_F, scales as K_F^2 or $n^{2/3}$. It rises faster than the gravitational pressure, so equilibrium is possible:

$$U_F \sim NE_F \sim N^{5/3}/V^{2/3}, \quad P_F = dU_F/dV \sim n^{5/3} \tag{2.14}$$

As the density rises with the contraction of the star, the electrons become relativistic, and their energy scales only linearly with the momentum rather than as the square. At this point, the electrons cannot halt the gravitational contraction. The electrons are pushed into the protons by the process $e^- + p \rightarrow n + \nu_e$, neutrinos are emitted, and the resulting neutrons continue to resist the contraction while they are non-relativistic. For a total number of nucleons N, a stable radius R_n exists in the balance of gravity with pressure $\sim 1/V^{4/3}$ and the neutron Fermi pressure, $1/V^{5/3}$:

$$R_n = (81n^2/16)^{1/3}\hbar^2 N^{-1/3}/Gm_n^3 \qquad (2.15)$$

The velocity of particles near the top of the Fermi sea is $\sim \hbar K_F/m$. If the contraction continues, the neutrons will also become relativistic:

$$\beta \sim (\pi\hbar/mc)(3^{1/6})n^{1/3}/\sqrt{2} \qquad (2.16)$$

The gravitational energy scales as N^2, while the fermionic energy scales more weakly with N. When the neutrons become relativistic, $p_F = \hbar c n^{4/3}$. A sufficiently massive star, with mass $\sim$ a solar mass, cannot stop the gravitational collapse. The following code graphs the Fermi pressure of electrons and neutrons as a function of R for a star with a user-chosen mass. At the higher masses, it is clear that collapse cannot be avoided. The points are shown only for the regime of NR fermions. The Schwarzschild radius for the Sun, a black star, is also shown for reference as is the current data point for the Sun, red triangle. The Schwarzschild radius is the radius where the gravitational force becomes strong and where general relativistic effects are expected to become important, since the potential energy is $\sim Mc^2$ at that radius.

The solar radius where collapse cannot be halted is a few kilometers. It seems that even the neutrons cannot halt a gravitational collapse. At 5 solar masses, the electron and nucleon pressure are always below the gravitational pressure. At 1/2 a solar mass, the fermions rise above the gravitational pressure and collapse can be averted. The conclusion is that Fermi pressure is insufficient to prevent gravitational collapse of a star with a mass of a few solar masses. These collapsed stars, black holes, are a prediction of strong GR since upon collapse the $\sim$ size of the star becomes the Schwarzschild radius.

Note that black holes are therefore thought to have evolved late in the history of the Universe as stars form from in collapse of gas clouds, ignite and age, finally exhausting their fusible fuel. These processes can take billions of years, for example, the lifetime of the Sun. This view of how black holes form is in some tension with new observations of super-massive black holes existing quite early, less than 1 Gyr since the Big Bang (BB), compared to the present age of ~13.8 Gyr.

The following code graphs the Fermi pressure of electrons and neutrons as a function of R for a star with a user-chosen mass and plots the results as in Fig. 2.5. At the higher masses, it is clear that collapse cannot be avoided.

```
% constants, MKS
me = 9.1e-31;
mn = 1.69e-27;
hbar = 1.11e-34;
c = 3.0e8;
G = 6.67e-11;
k = 1.38e-23;
% solar parameters
rsun = 6.96e8 ;   % solar radius , mass and mean density
msun = 2.0e30;
rhosun = (msun .*3.0) ./(4.0 .*pi .*rsun .^3) ; % assumed uniform
Psun = 2.6e16; % solar core pressure in Nt/m^2
M = 4.7; % Mass in solar mass units
Rs = (2.0 .*G .*M) ./(c .^2)    % Schwartzchild radius (m)
```

```
Rs = 1.3933e + 04
```

```
N = M ./mn; % number of nucleons
rho = M ./V; % mass density - uniform
n = rho ./mn; % number density - uniform, all H, Z/A = 1
% evaluate gravitational and e, n  pressure
Pg = (3 .*G .*M ^2)./(8 .*pi .*R .^4); % core pressure - const
density
Re = (Re .*8 .*pi) ./(3 .*G .*M .*M) % radius where e become
relativistic
```

```
Re = 1.8623e+06
```

```
Rn = (Rn .*8 .*pi) ./(3 .*G .*M .*M) % radius where n become
relativistic
```

```
Rn = 1.0028e+03
```

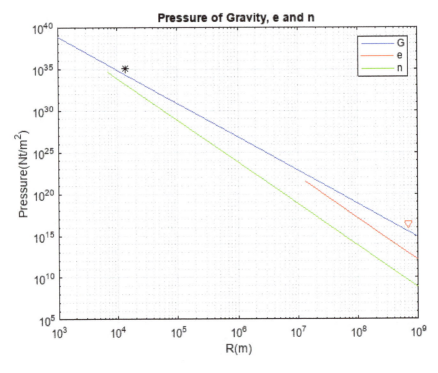

Figure 2.5:   Solar pressure as a function of radius for gravitational pressure and Fermi pressure of electrons and neutrons. The black star is the Schwarzschild radius for the chosen mass and the red triangle is the actual solar radius.

2.4   White Dwarf Prediction

It is possible that the mass of a star is small enough that it can be stabilized against gravitational contraction by the Fermi pressure of the electrons or, as a last resort, the neutrons. As fusion progresses, heavier nuclei are fused. However, beyond the iron nucleus, no further exothermic reactions are possible and the star will cool. All the heavier elements come from supernova explosions, not ordinary fusion reactions. In what follows, the assumption is that the star is to be stabilized by Fermi pressure. A simple model was previously made by looking at electron and neutron Fermi pressure and cutting off when the fermions became relativistic. In that simplest model, a uniform density star was assumed, and a mass less than about 2 solar masses could be stabilized. In this script, a smooth interpolation for NR to UR is made in an effort to make a somewhat more realistic model. The Fermi momentum is p_F and the interpolation function, γ_F goes smoothly from NR Fermi momentum to UR with x^2 to x behavior. The electron-to-nucleon ratio in the star is Ye ~ 0.5 but < 0.5 would apply assuming iron nuclei as the fusion endpoint:

$$n_o = (m_e c/\hbar)^3/3\pi^2$$
$$x = p_F/(m_e c) = (n/n_o)^{1/3}$$
$$\gamma_F \sim x^2/\sqrt{1+x^2} \tag{2.17}$$
$$\rho = (m_p n_o)/Y_e$$

The numerical equations as the electrons become relativistic are solved using "ode45" and are

$$\mathrm{d}M(r)/\mathrm{d}r \sim r^2 \rho(r)$$
$$d\rho(r)/dr \sim [M(r)\rho(r)]/(\gamma_F r^2) \tag{2.18}$$

The results extend the nucleon pressures a bit more realistically. Nevertheless, an iron star, with no radiation pressure cannot be stabilized, in this simple model, if its mass is greater than ~ 2.0 solar masses. Such a star must collapse into a singularity, a "black hole". The exact mass depends on the stellar composition. In addition, in GR the pressure is not the classical value assumed so far. More detailed models have been made, beginning with the first mass limit, the Chandrasekhar limit, of 1.4 solar masses. Present models find the limit in the range of 2–3 solar masses. However, it seems clear that a black hole is expected to be formed when stars a few times the mass of the Sun burn out. This expectation is borne out by astronomical observations of radiation from accretion disks as matter falls into a black hole and by lensing of background stars by intervening black holes.

```
% Model of White Dwarf - Pressure from e, Mass from Fe nuclei
% constants, MKS. UR to SR interpolation
% e supply pressure, Fermi gas at T = 0
% nuclei supply mass ~ at rest
% # density in #/m^3 when pf = mec (UR/NR boundary)
no = ((me .*c .*sqrt(2))./(pi .*hbar .*(3 .^0.1666))) .^3
```

```
no = 9.2587e+35
% star cannot be stable at much larger e densities
% overall charge neutral -> e number density => star mass density
rho = (mp .*no) ./Ye % nucleon mass density at no for e
```

rho = 3.3323e+09

```
% at this density, e become relativistic and cannot support G
% scale for radius and mass at ne = no in constant denstiy
approximation
Rosq = ((no .^1.666) .*(hbar .^2)) ./(me .*G .*rho .*rho);
Ro = sqrt(Rosq) .*2 ;
Mo = (4 .*pi .*rho .* (Ro .^3)) ./3;
Rsun = 7 .* 10 .^8 ;% sun radius
Msun = 2 .* 10 .^30 ;% sun mass
Rscale = Ro ./Rsun
```

Rscale = 0.0105

```
Mscale = Mo ./Msun
```

Mscale = 2.8046

```
% Rs = (2 .*G .*Mo) ./c .^2; % Schwarzschild radius
max(MM) % maximum mass for electron stabilized, try neutrons too
```

ans = 1.9707

```
function dy = dwarf(t,y)
    % t is the parameter = radius here
    dy = zeros(2,1);
    x = y(1) .^0.3333; % x = n/no
    % function to smoothly move from NR to UR
    gam = x .^2 ./(3 .*sqrt(1 + x .^2));
    dy(1) = - (y(2) .*y(1)) ./(gam .*t .*t);
    dy(2) = y(1) .*t .*t;
```

With the smooth momentum interpolation from NR to UR behavior, the maximum neutron stabilized white dwarf mass is still only about 2 solar masses contained in a radius of about 2.4% of the initial solar radius, as seen in Fig. 2.6. It appears that collapse into a singularity, a black hole,

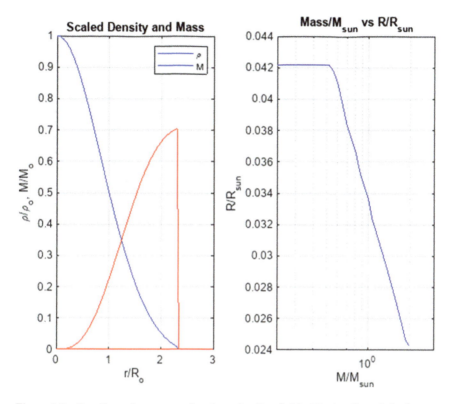

Figure 2.6: Density and mass as a function of radius (left). Final radius of the Sun as a function of mass (right), with instability at ~2 solar masses.

is impossible to evade if the physics of solar processes is indeed correct. Therefore, black holes should be plentiful in the Universe. The radial path of light in an SR space is $d^2s = 0 = (cdt)^2 - d^2r$ and all events can be seen if you wait long enough. However, in a GR Schwarzschild space light radial geodesics are $ds^2 = 0$, $dr/cdt = (1 - r/R_s)$ and the photon velocity approaches zero at the singularity. Such behavior is seen in Fig. 2.7 below where the black hole is "seen" through the light emission of particles in the accretion disk, but nothing inside the horizon. Indeed, this is a test of strong GR that a "horizon" exists from which nothing can escape and into which nothing can be seen. Event horizons are important in GR and specifically in cosmology as is explored later.

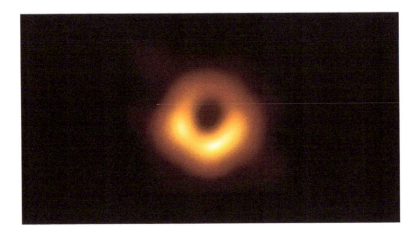

Figure 2.7:   Photo of a black hole with the accretion disc seen at radii $> Rs$.

2.5   Solar Nucleosynthesis

Observing the Sun, and looking at the spectral lines, it seems the Sun is largely hydrogen, with very little deuterium and about 26% helium. The solar composition can be explained if it is postulated that there was a Hot Big Bang (HBB) and that the very early composition of the Universe was a mixture of all the particles of the particle physics Standard Model (SMPP). In the SMPP, shown in Fig. 2.8, there are three doublets of strongly interacting quarks and three of weakly interacting leptons in the matter sector of the SMPP. The force carriers are photons (electromagnetism), W and Z bosons (electroweak), and gluons (strong). The Higgs boson has the quantum numbers of the vacuum and imparts mass to all the other particles through its vacuum expectation value (VEV).

The earliest times in a cooling Universe that are understood by particle physicists are at the TeV (10^9 eV) energy scale. There would be a photon sea with a small admixture of quarks, leptons, weak interaction force carriers, W and Z, and Higgs bosons. As cooling progresses, the quarks bind together to form the lowest mass protons and neutrons. The lightest leptons, the electrons, remain along with the weakly interacting neutrinos. Known physics should be able to predict the behavior of the Universe from this mass scale down to the present and to extrapolate to predict the future. One crucial input is that there is matter; all the anti-matter annihilations left some small excess of matter, for reasons that are not yet understood.

This model is now complete, with no indication of new physics at higher energies. Indeed, higher energies are not yet available in the laboratory.

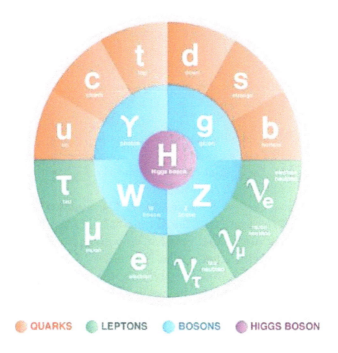

Figure 2.8: The standard model of particle physics.

That being the case, a closer look at the available data seems to be called for. Having successfully modeled the Sun, one can assume that the physics of stars is reasonably well known in the equilibrium state. However, it is necessary to understand the detailed composition of different nuclei in the Sun. In order to do that, it is necessary to understand the initial creation of the nuclei. The basic assumption is that there was an HBB, a fireball of photons with a small admixture of matter, electrons, neutrinos, and nucleons. The temperature of the fireball was, at short times after the HBB, so high that only the lowest mass particles existed. Any more complex bound states of these objects would be broken apart by the bath of high-energy photons.

For now define a time zero to be the HBB time. The Universe expands with a parameter H, the Hubble parameter, with dimensions of inverse time, such that $1/H$ is $\sim$ the age of the Universe. Taking the current measured value of H, H_o, the time is $\sim$14 billion years. The Sun can be expected to burn for billions of years after its formation, as was already seen. The Universe cooled in the expansion until more complex objects could exist without being continuously broken apart by the energetic photons. The strongest force, the nuclear force, has the largest binding energies. Due to this fact, the start of the history of the Universe begins with the formation of nuclei. This process happened within "the first three minutes". Numerically, at early times, $T\sqrt{t} = 10^{10}$ for T in K^o and t in s. At energies of a few MeV, only the lightest particles in the SMBP exist as a plasma of p, n, γ, e, ν. The observed present number density of cosmic microwave background (CMB), photons is $n_\gamma \sim 4.11 \times 10^8 \gamma/m^3$ with an energy density of $\rho_{\mathrm{CMB}} = 2.6 \times 10^{-4} \mathrm{GeV}/m^3$. They define the photon entropy. A reaction stays in thernal equilibrium, approximately, as long as the reaction rate exceeds the Hubble expansion, $\Gamma > H$, otherwise the reaction products are diluted by the spatial expansion and the reaction cannot remain in equilibrium.

Neutrinos decouple at a T about 1 MeV. The weak interactions couple with an effective constant, G_F, the Fermi constant, which is $\sim \alpha_W/M_W^2$ at low energies because of the massive W boson propagator which has mass M_W. The neutrino cross-section is of order $G_F^2 T^2$ and the number density goes as T^3. The mean free path then goes as $1/T^5, \Gamma \sim G_F^2 T^2$, while yields an estimated freeze out when $\Gamma/H \sim 1 \sim \alpha_W^2 T^3/M_p M_W^4$ or a temperature of a few MeV since the W boson mass is $\sim$80GeV with $G_F = 1.17 \times 10^{-4}/\mathrm{GeV}^2$.

A light nucleus has a typical binding energy of about 8 MeV per nucleon. There are so many photons that nuclei cannot form above a photon temperature of about 1 MeV since they would be broken up by the photons in the high-energy tail of the Planck energy distribution. For a temperature $T \sim 1$ MeV, only nucleons are in thermal equilibrium with the photons. At this temperature, the neutrinos decouple and propagate independently. Detailed estimates of the freeze-out of neutrinos yield an estimate of three neutrino generations, in agreement with the SMPP. For the nucleons, a precise calculation can be made because the masses and cross-sections are all well measured in the laboratory. At a temperature of $\sim$0.8 MeV, the neutrons and protons decouple from the photons, with a relative abundance given by their mass difference, $n_p/n_n = e^{Q/\mathrm{kT}}$, $Q = m_n - m_p = 1.293$ MeV. The neutrons then decay and are also taken up into the formation of nuclei. The mass fraction of a species, X, is the abundance relative to all nucleons, which for neutrons is $X_n = n_{nl}(n_n + n_p)$. The Boltzmann transport equation for the neutrons is driven by the n-p reaction rate, Γ_{np}, $p + \nu + e^- \longleftrightarrow n$:

$$\mathrm{d}X_n/\mathrm{dt} = \Gamma_{np}[(1 - X_n)e^{-Qt/T} - X_n]. \tag{2.19}$$

These types of equations will be used extensively in later cosmological discussions. They are used to explore the evolution of the constituents of the Universe as it evolves in time under GR Hubble expansion. The n-p reaction is a weak interaction process, which scales as T^5. At the n freeze-out temperature of ~0.8 MeV, the n to p ratio is about n_p/n_n. After this time, the n decay and the n fraction at the time of helium formation is only ~11% when T is ~0.3 MeV, and all the n remaining at that time are taken up into He. Since there are $2n$ in each He nucleus, an estimate of the primordial helium abundance with respect to all nucleons is ~22%. A slightly more detailed estimate of the primordial helium abundance can be made. A simple model, called nuclear statistical equilibrium (NSE), is used to estimate the abundance of light nuclei. The nucleon-to-photon ratio is taken to be $\eta = 5.5 \times 10^{-10}$ from knowledge of the present CMB. There is a competition in nuclear formation between the binding energy B_A and the photons to form a nucleus with Z protons and A-Z neutrons. The small value of η (photon entropy) means that nuclei freeze out at temperatures well below their binding energies. At a given temperature, T, Z protons and $(A$-$Z)$ neutrons come together and bind a nucleus A, with an energy B_A:

$$X_A \sim T^{3(A-1)/2}\eta^{(A-1)}X_p^Z X_n^{(A-Z)}e^{B_A/T}. \qquad (2.20)$$

Specifically, n and p form deuterium D. The abundance of D is small because its binding energy is only 2.2 MeV which makes it very sensitive to η, by way of the process of deuterium breakup, $n + p \longleftrightarrow D + \gamma$. The protons are taken up in He formation as are the neutrons. The simplified set of approximate Boltzmann equations to be solved are.

$$\begin{aligned}
&X_A = n_A A/n_N, \quad 1 = X_p + X_n + X_D + X_{\text{He}}\\
&X_n = X_p e^{-Q/T},\\
&X_D \sim T^{3/2}\eta e^{(B_D/T)}X_p X_n\\
&X_{\text{He}} \sim T^{9/2}\eta^3 e^{(B_{\text{He}}/T)}X_p^2 X_n^2
\end{aligned} \qquad (2.21)$$

The n do not freeze out in this approximation nor do they decay. The p become dominant until taken up by He formation, which abundance rises very rapidly with falling T to about 22% at $T \sim 0.28$ MeV. The following script simply assumes $1/2n$ and $1/2p$ at high temperatures and then evolves to lower T. Using known SMPP physics and an assumption of a HBB, the primordial nuclear abundance can be predicted and checked against spectroscopic data. The user can choose the size of the photon entropy and see the effect on the helium abundance, as shown in Fig. 2.9. The HBB model appears to describe the elemental makeup of primordial stars at energies ~ nuclear binding energies of ~MeV.

```
% evaluate mass fractions in Nuclear Statistical Equilibrium,
Follows Kolb - Turner text
% Light Elements in NSE ,   starting values, only p and n at high T
y1(1) = 0.5; % p
y2(1) = 0.5; % n
y3(1) = 0.0;   % D2
y4(1) = 0.0;   % He
Q = 1.293; % n-p mass difference
B2 = 2.22 ; % binding energy of D, He
B4 = 28.3 ; % main dependence is B4 vs photon entropy
mn = 938.0; % nucleon mass, MeV
% entropy - nucleon to photon ratio, photons from from CMB + v
eta = 7.15e-10;
fact = 0.9 ; % factor to scale entropy, 0.8 to 1.2
eta = eta .*fact;
% a simple numerical rundown of the system of equations
```

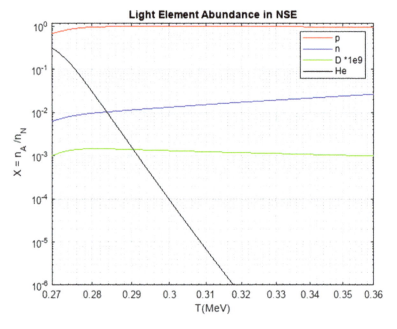

Figure 2.9:   Approximate abundance of light elements as a function of HBB temperature.

```
y4(193) % He abundance - approximate, rapid T and entropy dependence
```

```
ans = 0.2350
```

The plots are not correct in detail, more detailed model results are shown in Fig. 2.10. However, they show p take-up to feed He production and the rapid rise of He at T scales much less than the binding energies. They also show the very small D abundance and the large He abundance. Instead of the 22% He abundance expected, the Sun has 28%. Indeed, the Sun is not a first generation star. Heavier elements are expected to be blocked by the large He binding energy. Other, younger, stars are observed with abundances closer to 22%. Indeed, heavier elements up to iron are thought to arise in older stars that have used up the light elements to fuel the fusion reactions, or as supernova remnants which are accreted into second and later-generation stars. Elements heavier than iron can only arise from supernovae because iron has the peak binding energy per nucleon. The Sun has a helium mass fraction of 0.28. However, the Sun is not a first-generation star and so it does not possess the primordial helium fraction which can be extrapolated from observations and is in agreement with the nucleosynthesis model. Primordial X for He is $\sim$0.24. The observed fraction in the Sun is $\sim$0.28. However, the Sun has been burning p and making He and releasing 28.4 MeV per He nucleus. The Sun consists of about 10^{57} protons and emits about 10^{36} GeV/sec or $\sim 10^{54}$ GeV over a 10 Gyr lifetime, creating $\sim 3 \times 10^{55}$. He nuclear ashes. This means X for He is now $\sim$0.24 + 0.03, close to the observed value.

2.6 Higgs Vacuum Expectation Value (VEV)

A major recent particle physics advance was the discovery of the Higgs boson. This fundamental particle is a scalar, with the quantum numbers of the vacuum. It is the sole such particle of all the fundamental particles of the SMPP. This unique particle would have cosmological implications because it is an intrinsic part of the vacuum as is the postulated cosmological term, $\Lambda g_{\mu\nu}$, for DE. It therefore needs to be explored a bit more in depth. For a free scalar field ϕ, the Lagrange density is $l = \phi(p_\mu p^\mu - (mc^2)^2)\phi$, where p_μ is the 4 vector momentum and m is the mass. The basic form is familiar from the relationship $\varepsilon^2 = (cp)^2 + (mc^2)^2$ with an SR vector with components for energy ε and momentum, $\overrightarrow{p}$, $p_\mu = (\varepsilon, \overrightarrow{cp})$. The Higgs boson has self-interactions which are defined to have a potential, V, with quadratic and quartic dependence on the Higgs fields $V(\phi) = \mu^2 \phi^2 + \lambda \phi^4$, parameterized as a Taylor expansion in the fields. The field has the dimension of energy, while the potential has a dimension of energy to the fourth power. This potential has a minimum at a field value $\langle \phi \rangle^2 = -\mu^2/2\lambda$

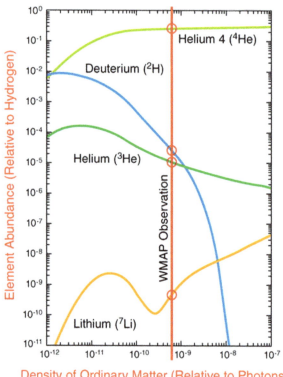

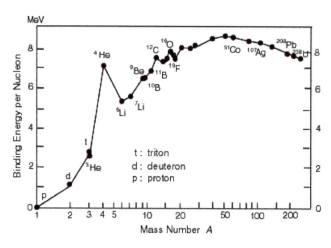

Figure 2.10: Abundance of light solar elements (top) and binding energies of nuclei (bottom). The observed abundances, red circles, and the HBB predictions are in very good agreement.

and the potential minimum is

$$V(\langle\phi\rangle) = \mu^2\langle\phi\rangle^2 + \lambda\langle\phi\rangle^4, \quad V_{\min} - \mu^4/4\lambda \tag{2.22}$$

The physical ground state Higgs boson appears when one expands about the potential minimum, which is the minimum of the true vacuum, $\phi = \langle\phi\rangle + \phi_H$. The potential is then

$$V(\phi_H) = -\lambda\langle\phi\rangle^4 + 4\lambda\langle\phi\rangle^2\phi_H^2 + 4\lambda\langle\phi\rangle\phi_H^3 + \lambda\phi_H^4 \tag{2.23}$$

These terms can be identified from left to right as the constant potential minimum, a Higgs mass term with mass squared $\lambda\langle\phi\rangle^2$, and the Higgs self-interactions involving 3 and 4 Higgs at an interaction vertex — triplet and quartic couplings with coupling constants $\sim\lambda$. Indeed, these predicted couplings are being very actively searched for at colliders, such as the Large Hadron Collider (LHC) at CERN. Such couplings for the electroweak force carriers, the W and Z bosons, have been observed already at the LHC, for example, with the observation of W pair production indicating triple W couplings. Evidence for triple W production has indicated the existence of quartic couplings in the SMPP. These predictions and their observation are confirmation of the SMPP. Presently, there is a program to search for the production of pairs of Higgs to see if the predicted strength of the triple Higgs vertex coupling is confirmed. The confirmation is likely to require an increase in the LHC luminosity or reaction rate as planned with an upgrade to the accelerator and detectors.

The potential is then defined by two parameters: μ and λ. One parameter is measured by using the boson masses, γ, g, W and Z where the photons and gluons are required to be massless. For the weak force carriers, W and Z, their mass is supplied by the Higgs since they couple to $\langle\phi\rangle$. The second parameter, λ, is measured once the Higgs particle mass is determined, which completes the determination of the Higgs potential, assuming no higher-order terms exist.

```
syms V u l ph ph2 pH
V = u^2*ph2 +l*(ph2^2) % potential
```

$$V = lph_2^2 + ph_2u^2$$

```
eqn = diff(V,ph2) == 0; % minimize
sol = solve(eqn,ph2) % phi^2 for min V
```

```
sol =
```

$$-\frac{u^2}{2l}$$

```
Vm = subs(V,ph2,sol) % V at min
```

Vm =

$$\frac{u^4}{2l}$$

Using the assumed potential, a minimum occurs at

$$\langle\phi\rangle^2 = -\mu^2/2\lambda., \quad V_{\min} = -\mu^4/4\lambda$$

$$\phi_H, \phi = \langle\phi\rangle + \phi_H \tag{2.24}$$

$$V = \mu^2 |\langle\phi\rangle + \phi_H|^2 + \lambda(\langle\phi\rangle + \phi_H)^4$$

$$M_H = \sqrt{\lambda}\langle\phi\rangle$$

Having found the minimum, one expands around it with the physical Higgs field. The result contains a mass term for the Higgs, quadratic in the Higgs field. The Higgs also induces a mass in the weak gauge bosons. The measured mass of those weak bosons determines the parameter, $\langle\phi\rangle$ to be 174 GeV. The remaining parameter is found with the discovery of the Higgs boson itself, at a mass of 125 GeV, which determines $\lambda = 0.50$ (Parenthetically, the SMPP has symmetry group structures, SU(3) × SU(2) × U(1) for the strong, electroweak, and electromagnetic interactions, and the present treatment has ignored this fact, which necessitates a tweak for the expression for the Higgs mass because of the group "Clebsch–Gordan" factors.) The user can choose the quartic coupling and see how that changes the shape of the potential, with an example displayed in Fig. 2.11.

```
% phiminsq= -mu*mu/(2*lambda)
% numerical constants
% phimin ~ 174 GeV
phimin = 174;
lamda = 0.88 ; % 0.88 is measured;
mu = -sqrt(2 .*lamda .*(phimin .^2))
```

mu = -220.0945

```
Vmin = -(mu .^4) ./(4 .*lamda)
```

Vmin = -7.3331e+08

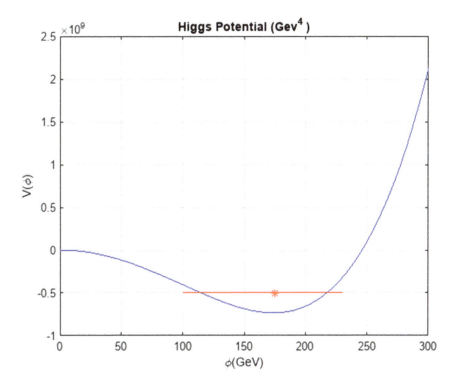

Figure 2.11: Plot of the Higgs potential as a function of the field. The minimum is the red dot and the oscillation about it is the schematic red line.

```
[MH,imin] = min(V); phi(imin)
```

```
ans = 172.7273
```

The VEV is a unique attribute in particle physics and induces a mass to all the other fundamental particles. At present, the SMPP agrees with all present data. However, it has many constants, masses, and couplings, which are not predicted but are experimental inputs. Therefore, there is no clear experimental path to explore. Higher energy accelerators may uncover a new and higher mass scale, but the LHC has not yet done so. Thus, this mass scale may require yet another increase in the energy of accelerators.

As for cosmology, there is a standard model, SMC, which also agrees with all the present data. However, ordinary matter is only ∼5% of all the mass of the Universe. There is dark matter (DM), with no SMPP candidates in sight and no experimental observations yet beyond that it gravitates. There is also DE, at present experimentally compatible with a cosmological constant. It is the majority energy of the Universe, ~70% and no SMPP candidates exist for it. The Higgs is an SM particle with vacuum quantum numbers, but it does not appear to gracefully be a candidate, although it shows that fundamental scalars exist. Hence, there are many, many open questions in physics, and an interesting future seems to be assured.

It is also potentially a problem that the Higgs potential has a non-zero minimum value, as seen in the plot above. The value of $V(\phi_H)$ is $-\lambda\langle\phi\rangle^4 \sim -M_H^2\langle\phi\rangle^2$ which has a value $\sim 10^8$ GeV^4 many orders of magnitude, ~ 54, larger than the value assigned to the cosmological DE, which implies that the Higgs vacuum is gravitationally inert. This issue is raised here only to be dropped to await further progress.

2.7   Gravitational Radiation

Accelerated quadrupole mass distributions, with an energy–momentum stress tensor, $T_{\mu\nu}$, are expected to radiate gravitational waves. The metric disturbances, $h_{\mu\nu}$, to the background metric follow a wave equation in the weak GR approximation, where $\eta_{\mu\nu}$ is the flat space diagonal Minkowski metric:

$$\frac{\partial^2}{\partial r^2}h_{\mu\nu} - \frac{\partial^2}{\partial(\mathrm{ct})^2}h_{\mu\nu} = -16\pi G(T_{\mu\nu} - (T/2)\eta_{\mu\nu}) \qquad (2.25)$$

In electromagnetism, an accelerated charge radiates. Mass is the GR "charge". Therefore, an accelerated mass should radiate. The mass defines the metric, so a gravitational wave should be a metric distortion. These effects are not normally obvious to say the least, so how does one establish that gravity waves exist? Since matter (and pressure) are the source terms for the metric in GR, extreme astronomical situations with distance scale near the Schwarzschild radius are places to look for gravitational radiation. A binary star or binary black hole system is a candidate source. It will lose energy by radiating gravitational waves and will in-spiral to near the Schwarzschild radius from an initial orbit. The situation is similar to the old problem that atoms are stable but cannot classically exist because they will radiate electromagnetic waves. That problem was solved by quantum mechanics, but a quantum theory of gravity is not yet available. In this situation, classical dynamics is applied up to the motion of the radius R_s, with an energy loss due to GR radiation. The scale of the maximum wave frequency for a solar mass source scales is in kHz. A collapse to a black hole is as extreme a situation as can be obtained and massive enough stars must collapse into such objects. Gravity does win in the end.

The radiation is quadrupole as opposed to the lowest order dipole electromagnetic radiation. There is no negative mass, so no dipole exists whereas charge has both negative and positive constituents. Therefore, the power radiated scales as ω^6 and not the electromagnetic fourth power. Any gravitational system can only radiate up the maximum power, $P_{\max} = 2c^5/5G$, which is $1.32 \times 10^{52} W$. This is approximately 10^{26} times larger than the solar luminosity, which allows it to be observed. For the in-spiraling binary, the power is $\sim$ the maximum power times $(R/R_s)^5$. Consider a binary star system rotating about the center of mass (CM), frame at $(0, 0)$ with masses $M_1, M_2 = M$ and distances $r_1, r_2 = R$. The user chooses the mass of the binary stars, which is essentially the only variable in the problem except for the initial radius. In the simplest case of equal mass stars the rotation frequency ω, power output dU_G/dt, and the reduction in the period $d\tau/dt$ as energy is lost are,

$$\omega^2 = GM/4R^3$$

$$dU_G/dt = (2c^5/5G)(R_s/2R)^5 \qquad (2.26)$$

$$d\tau/dt \sim (R_s/c\tau)^{5/.3} \sim (R_s/R)^{5/2}$$

The following script makes a "movie" of the classical in-spiral and then plots the "chirp" down to a separation of R_s with an example output shown in Fig. 2.12. The LIGO discovery data is shown in Fig. 2.13 for comparison. The dynamics is classical since a full GR solution does not exist. As a rule of thumb, large objects like whales make low-frequency sounds and vice versa. A solar scale mass has a Schwarzschild radius $\sim$km which implies radiation with a final wavelength, $\lambda \sim R_s$.

```
% Inspiraling Binary due to Grav Radiation, classical binary
dynamics, GR
% energy loss
G = 6.67e-11; % mks units,
Mo = 2.0e30; % solar mass,
c = 3.0e8;
mm = 5; % mass in Mo units (1,10)
M = mm .*Mo;
rors = 4.0; % ro in rs units (3,5)
rs = (2.0 .*G .*M) ./c .^2; % Scwarzchild radius
ro = rors *rs; % (m)
% find the initial orbital frequency and classical time to inspiral
```

```
wo = sqrt((G .*M) ./(4.0 .*(ro .^3)));
tc = (5.0 .*(ro .^4) .*(c .^5)) ./(32 .*((G .*M) .^3));
% time to rs
ts = tc - (5.0 .*(rs .^4) .*(c .^5)) ./(32 .*((G .*M) .^3));
ws = sqrt((G .*M) ./(4.0 .*(rs .^3)));
% Initial Orbital Frequency (Hz) = wo, Classical Inspiral Time to r
=0 (sec) = tc
% Schwarzchild Radius (m) = rs,  Max Frequency at rs = ws
% Time to Reach Schwarzchild Radius (sec)= ts
tss = ts .*1000 % inspiral time im msec
```

tss = 15.7486

```
wss = ws ./1000 % max frequency, at rs, in kHz
```

wss = 7.1559

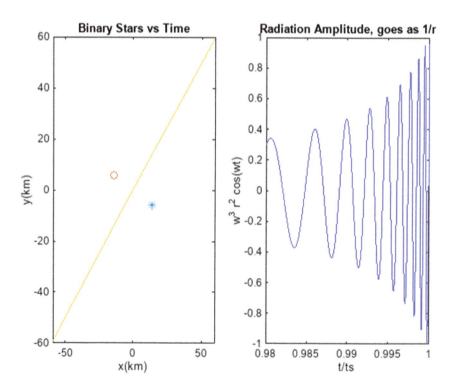

Figure 2.12: Orbit of the binary system (left) and the radiated power vs time (right).

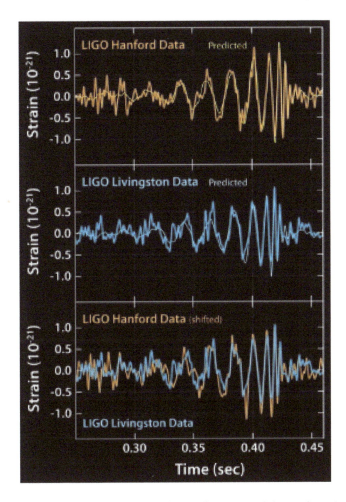

Figure 2.13: Data on the deformations for the first reported detected gravity wave.

The shape of the radiated power and the observed detected gravitational wave are shown above. It is clear that the present simplified classical model for the radiated power gives an adequate first approximation to the data. The recent experimental discovery of gravity waves confirms GR radiation theory. The quadrupole "antennas" for solar mass radiation need to be of the scales of the solar Schwarzschild radius $\sim$km which is indeed the size of the LIGO interferometer. Note that this is a confirmation of strong GR where the scales are near those of the singularities, $R \sim R_s$.

To summarize Section 2, the Sun is understood in some detail with the advent of nuclear physics. The quantum Fermi pressure is unable to prevent the collapse of massive stars into singularities. The measurement of primordial Helium supports an HBB model of the early Universe containing matter and radiation. The CMB photons provide the needed HBB entropy as a crucial input. A fundamental scaler boson, the Higgs, exists and has the quantum numbers of the vacuum. Disturbances of the metric manifest themselves as gravitational waves which are detectable. These, and other, known and relevant phenomena will be used in the ensuing discussion of the standard model of cosmology (SMC).

Chapter 3

The Cosmic Metric

"Look up at the stars and not down at your feet. Try to make sense of what you see, and wonder about what makes the universe exist. Be curious."
— Stephen Hawking

"When you look at the stars your eyes are time machines."
— Waqas Rabbari

A fundamental fact about the Universe is that it is expanding as determined by Doppler shifts in spectral lines of the light from distant stars. The Hubble parameter quantifies the expansion and has dimensions of inverse time as it is defined to be the fractional time rate of the expansion of physical distances. The Hubble distance, $D_H = c/H_o$, is then the approximate size of the Universe, assuming a HBB and that the current Hubble parameter has been constant for all time. The subscript o is used in the text to denote the present value of a parameter, as is conventional for cosmology. There is some question as to the exact value of the scale h at the 10% level. In this text, h is fixed to the value indicated in Eq. (3.1), but equations with h left unspecified are often encountered in the literature by including a factor h, which is assigned a value 0.68 in this text:

$$H_o = 100\, h * \mathrm{km}/(\sec * \mathrm{Mpc})$$
$$1\ pc = 3.08 \times 10^{16}\mathrm{m} \tag{3.1}$$
$$D_H = c/H_o = 4440\ \mathrm{Mpc}, \quad h = 0.68$$

A parsec is a customary unit in cosmology, but in this text, the international scientific MKS units will normally be used for mass, length, and time. For reference, 1 Mpc $= 3.08 \times 10^{22}$ m. In these units, $c = 306$ Mpc/Gyr so that in 13.8 Gyr, light in a flat space has traveled 4220 Mpc. The Hubble parameter can be cast as a length, the Hubble distance (D_H) as an energy, $\hbar H$, or as a time, $1/H$. The time since the HBB is $\sim 1/H_o$ or ~ 14.5 Gyr. Old clusters of galaxies have an age ~ 10 Gyr which roughly agrees with terrestrial U decays which have Gyr lifetimes and both confirm, approximately, the age of the Universe.

The reader should not worry about these equivalent representations and become comfortable with them. For example, $H_o = 1.46 \times 10^{-33}$ eV, $T_o = 2.34 \times 10^{-4}$ eV $= 2.726$ K°. Just to get an idea of the size of the visible Universe, take the volume $\sim D_H^3$ with an average energy density, 5.6 GeV/m^3. The mass is $\sim 10^{78}$ GeV consisting of $\sim 10^{13}$ or 10 trillion galaxies which in turn each contain $\sim 10^{21}$ stars each of about 1 solar mass. In turn, there is on average ~ 1 galaxy per cubic Mpc. It's a big Universe out there. Indeed, the Robertson–Walker (R–W) metric smooths out the galaxies over a scale of ~ 100 Mpc, averaging over ~ 1 million galaxies.

Newton's constant G will also be used, but mass, length, and time units can also be derived from G by multiplying or dividing by c (units length/ time) and by $\hbar$ (units energy * time) or $\hbar c$ (units momentum * length). These "Planck units" for mass, length, and time, where gravity is a strong force, are

$$M_p = \sqrt{\hbar c/G} \sim 1.2 \times 10^{19}\,\text{GeV}$$
$$L_p = \sqrt{\hbar G/c^3} \sim 1.6 \times 10^{-35}\,\text{m} \tag{3.2}$$
$$t_p = L_p/c \sim 5 \times 10^{-44}\,\text{s}$$

For example, G can be written as $G = 1.17 \times 10^{-5}(\hbar c)^3/\text{GeV}^2 = \hbar c/M_p^2$. The advantage of using Planck units is that in these units, $c = G = k = \hbar = 1$. The reader should not be intimidated by this convention. All the scripts used in the text report numeric results in MKS units. One can also simply convert Planck units to MKS units by repeated application of the constants c, k and $\hbar$. For example, in the text, the mass density of a scalar field ϕ is said to have a term $\rho = (d\phi/dt)^2$. Fields like the Higgs have dimensions of energy. The appropriate factors to multiply can most easily be done in steps as $(1/\hbar c)(1/c^2)$. Dimensionless quantities are also often used, such as fractional inhomogeneity, conformal time, and comoving distance. The electron volt will be the typical energy unit used here. The literature for cosmology uses these units, so the reader needs to become comfortable with them. In particular, G will be dropped in what follows, replaced by the Planck mass since that mass scale is where gravity becomes strong.

3.1 Hubble Visualization

As discovered by Hubble, using the red shifts of spectral lines, space is expanding, with a velocity which increases with distance, R, $v \sim H_o R$. A very schematic model of the expansion of space itself, where every free fall, or comoving, observer sees neighboring points moving away, is shown in Fig. 3.1. The user chooses the point of view of the observer on a chosen grid point. Coordinates of the metric used in what follows are comoving coordinates, r, in free fall with respect to the expansion and are constant, while physical distances, $R = a(t)r$, expand in time as tracked by an overall scale factor $a(t)$, with a present value $a_o = a(t_o) = 1.4 \times 10^{26}$ m. Time $= 0$ is the time of the "Big Bang", while t_o is the present time so that $a(0) = 0$, $H_o = H(t_o)$.

```
% movie to illustrate expansion - Hubble. Pick observer location and
  plot w.r.t. that
% user chooses location of observer along x
icen = 6;
```

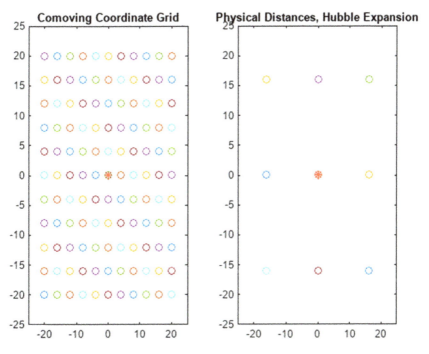

Figure 3.1: Grid of comoving distance coordinates (left) and location of physical coordinates after a period of expansion (right).

This, and other movies, can be played back at user choosen speeds using the controls at the bottom of the figure.

3.2 Radiation and Matter, Power Laws

Something happened about 13.8 billion years ago, which we now call the Big Bang (BB), or hot Big Bang (HBB). This event occurred with extreme temperatures and as the Universe expanded, it cooled. As of today, a sea of photons is observed, the cosmic microwave background (CMB), which is very uniform spatially and has a black body equilibrium temperature distribution. Matter is also observed as clumped aggregates: suns and galaxies and clusters of galaxies. In large enough volumes, of size ~100 Mpc, it is of uniform density and is isotropic. This conclusion is substantiated by telescopic surveys of millions of galaxies and by the CMB itself. From these observations, an attempt can be made to extrapolate backward toward the initial event. Since gravity always "wins", the extrapolation must use a GR formulation. The language of cosmology is GR.

The metric is perhaps the ultimate "spherical cow", depending only on a single parameter, $a(t)$. The Universe is assumed to be homogeneous and isotropic with comoving coordinates, r, θ, ϕ. Comoving distance is scaled by a parameter $a(t)$ to physical distance, R. The time is labeled by a Universal clock time, t, where time is the time on clocks of comoving observers in free fall. It is also known that all galaxies appear to be moving away from us, the Hubble discovery of Doppler shifts in the spectra of distant galaxies. Since we are not privileged observers, it must be that space itself is expanding, which is possible in GR where the metric is sourced by the energy content. The history of the Universe is defined by the time evolution of a single parameter, the scale parameter $a(t)$, which has the dimensions of length. This metric is called the R–W metric. Under these assumptions the Universe and it's history is contained in $a(t)$:

$$
g_{\mathrm{uv}} = \begin{bmatrix} 1 & 0 & 0 & 0 \\ 0 & -a^2 & 0 & 0 \\ 0 & 0 & -a^2 & 0 \\ 0 & 0 & 0 & -a^2 \end{bmatrix}
\tag{3.3}
$$

$$
ds^2 = (cdt)^2 - a^2(t)[d^2(r) + r^2 d^2\Omega]
$$

The metric is defined by a simplified model, not here derived from the GR field equations. However, the elements of the metric can be simply related to the stress–tensor components, matter and pressure. With the assumed R–W metric, the geodesic equations for that metric can be derived. The dynamics of the Universe, the R–W metric, is driven by the energy content as expected in GR. The expansion of the Universe is defined by the Hubble parameter $H = \left(\frac{d}{dt}a\right)/a$ and the density ρ is the total energy density of matter/energy of all types. Note that researchers in this field often use cgs units preferentially. Units with $c = G = \hbar = k_B = 1$ are also often encountered, so the reader of texts should be aware of that complication. In this text, MKS units are consistently used in the final numerical results, although simplified equations may occasionally appear if it is convenient.

An heuristic Newtonian argument can be made for the geodesic equations. Consider a sphere of uniform density and radius $R(t)$ with a fixed mass M. A mass at this radius need only track the interior mass as is well known classically. The acceleration of that mass is $-GM/R^2$. A first integral yields the energy conservation equation, with integration constant C. Picking C of zero and defining $R = a(t)r$, the resulting equation for the scale factor assuming constant density, $M = (4\pi/3)\rho R^3$, which is the Friedmann equation, Eq. (3.4), if the density is generalized to be the relativistic energy density from all sources. This "derivation" illustrates the close relationship of the Friedmann equation and classical energy conservation. In fact, the classical analog of the critical density is "escape velocity" where an object may escape, expand forever, if sufficiently energetic to be able to avoid gravitational "collapse".

The energy density has a time dependence determined by the equation of state obeyed by the particular type of energy. A homogeneous Universe is all at the same temperature (the CMB) so that there is no net heat flow. In classical thermodynamics, an adiabatic process has no net heat flow, dQ, so that $dQ = 0 = dE + pdV$. The entropy change, dS, is also zero since $dS = dQ/T$. The basic relationship has to do with the change of energy within a physical volume V which is proportional to a comoving volume times the cube of the scale factor $a(t)$. This change, $d(\rho a^3(t))$, is due to the pressure, p, during expansion, $-pdV$. Differentiating the adiabatic equation, $d(\rho a^3(t)) + pdV = 0$, with $V \sim a^3(t)$, results in $d\rho/dt + 3(da/dt)(p + \rho)/a = 0 = d\rho/dt + 3H(p + \rho)$.

The specific geodesic equations for the R–W metric are called the Friedmann equations. The expressions for the first and second time derivatives of $a(t)$ are given with contributions, in general, for matter, radiation, dark energy (DE), chosen for now to be a cosmological constant Λ, and a possible curvature contribution with coefficient κ. The velocity of the Universe is driven by the familiar energy sources, ordinary matter and radiation with the addition of dark matter (DM). Another contribution to H is a possible curvature term, $-\kappa c^2/a^2$, as shown, but for a flat Universe, $\kappa = 0$, is almost always assumed in what follows. This assumption is made essentially because the data lead to that conclusion within present accuracy. In the past, there was a question whether the Universe was closed or open and whether it would expand forever or decelerate and collapse. Those questions have been subsumed by the observation of the extreme uniformity of the CMB and the postulate of "inflation" which was made in order to explain that, and other, data:

$$\left(\left(\frac{d}{dt}a\right)\bigg/a\right)^2 = H^2 = (8\pi\rho/M_p^2 + \Lambda c^2)/3 - \kappa c^2/a^2 \qquad (3.4)$$

The fluid flow equation for the evolution of the energy density in an expanding space is given in Eq. (3.5). This is the classical continuity equation for fluid flow with the added term in H which shows the loss in density due to the spatial expansion which acts like a cosmic "friction" term. It is also notable that in GR pressure is a source as is mass/energy as was already mentioned. Using these equations, the acceleration of the scale factor can then then derived. Note that the curvature does not appear as a source of acceleration. The velocity equation is the GR version of energy conservation, while the acceleration equation is the GR version of the Poisson equation for the R–W metric with the addition of a cosmological term proportional to Λ. Note that matter and pressure both cause deceleration of $a(t)$. A positive cosmological parameter will, however, cause acceleration of $a(t)$. Clearly, for both *MD* and *RD*, the acceleration is negative; an expanding Universe dominated by the energy density of matter or radiation will decelerate. The acceleration in a matter-dominated phase, goes as $1/a^2$, while in a radiation dominated phase, it goes as $1/a^3$. In either case, the deceleration is large at early times when $a(t)$ is small and then slows as the scale factor $a(t)$ grows:

$$d\rho/dt + 3H(\rho + p/c^2) = 0$$

$$\left(\frac{d^2}{dt^2}a\right)\bigg/a = -(4\pi/3M_p^2)(\rho + 3p/c^2) + \Lambda c^2/3 \qquad (3.5)$$

A critical density, ρ_c, is defined in Eq. (3.6) and densities are scaled to it; $\Omega(t) = \rho(t)/\rho_c$. Deviations from a total of 1 are driven by the curvature, $|\Omega(t) - 1| = \kappa/[H(t)a(t)]^2$. The measured value of the critical density is $\rho_c \sim 10^{11}$ solar masses per Mpc^3. Normally, in this text, it is assumed that the Universe is flat, $\Omega_m + \Omega_\gamma + \Omega_\Lambda = 1$ which is addressed later when inflation is discussed. The critical density and the contributions of different types of energies to that density are explained later. For now, it is important to note that the energy in the early Universe has contributions in increasing importance from radiation, ordinary matter, DM, and DE. The present value of a quantity is denoted by a o subscript. The Ω values quoted also refer to the present value and the subscript is omitted. For DE, $\Omega_\Lambda = c^2\Lambda/(3H_o^2)$. The program is to take current observations and attempt to extrapolate back in time toward the HBB. It is to be noted that the model which is required to fit the data, the SMC contains only about 5% by mass of objects known to the SMPP, with values shown in Table 3.1:

$$\rho_c = 3(H_o)^2/8\pi G = 3(H_o M_p)^2/8\pi, \quad \{H(t)/H_o\}^2 = \rho/\rho_c = \Omega(t)$$
$$\rho_c = (4.5 \times 10^{-47}\,\text{GeV})^4$$
(3.6)

The Schwarzschild radius for a point mass M is $R_s = 2M/M_p^2$. In a very simple minded extrapolation, consider a constant density dust ball with the critical density ρ_c. The mass inside R_s is evaluated in flat space and yields a radius $R = M_p\sqrt{3/(8\pi\rho_c)} = c/H_o$. The Schwarzschild radius is the radius from which no particle can escape, and it is amusing that for a classical dust ball, it is D_H, the present Hubble distance.

Gravity is ordinarily a very weak force compared to electromagnetism. We can overcome the whole attraction of the Earth with a small bar magnet, for example. A mass that sets the scale when gravity is strong is defined in terms of a "fine structure constant" for gravity. Numerically, the Planck

Table 3.1: Critical densities of the SMC as presently determined.

Quantity	Present value
ρ_c — critical density (GeV/m^3)	5.6
Ω_m — matter fraction (dark matter dominant)	0.315
Ω_b — baryon fraction	0.05
Ω_γ — radiation fraction	4.64×10^{-5}
Ω_Λ — dark energy or vacuum fraction	0.685

mass, M_P, is 1.2×10^{19} GeV. This enormous mass scale, $\sim 10^{16}$ times the present upper experimental energy scale of the SMPP, indicates when gravity becomes dominant. It is not the Schwarzschild radius of a specific mass point but the analog of the fine structure constant. There is no dimensionless constant, such as exists for electromagnetism, for gravity, which means that gravity cannot be kept under quantum field theory theoretical control at all energies, and there is, at present, no quantum field theory of gravity. Since gravity is strong at Plank mass scales, quantum effects may be expected to appear then. Indeed, quantum fluctuations will be seen to be critical elements of the theory of inflation:

$$\alpha = e^2/\hbar c \sim 1/137$$

$$\alpha_G = GM_p^2/\hbar c, \quad \alpha_G = 1, \quad M_P = \sqrt{\hbar c/G}$$

(3.7)

With an additional equation of state for the fluid, the pressure term can be converted to mass density, $p/c^2 = \omega\rho$. The acceleration of $a(t)$ can then be found. For matter, $\omega \sim \langle\beta^2\rangle/3 \sim 0$, the Maxwell Boltzmann result, while for radiation, $\omega = 1/3$, since $\beta = 1$ and the energy density of radiation scales as the fourth power of the volume. Note that both matter and radiation cause a cosmic deceleration. A dimensionless deceleration parameter is $[a(d^2a/d^2t)]/(da/dt)^2$. Acceleration of $a(t)$ occurs only if a "dark energy" or a negative, repulsive Λ term exists. If not, then the age of the Universe is $>1/H_o$ because of the deceleration. Using the acceleration equation and ρ, p the power-law behavior for matter and radiation is easily derived. These results are also presented in Appendix B. The deceleration parameter for power-law behavior $a(t) \sim t^n$ is $q = -(1-n)/n$. There is deceleration for both matter and radiation. However, DE as defined accelerates. Note that the sign of a cosmological term could *a priori* be of either sign, but here the sign is chosen by the observation of DE. The three sources are labeled as γ, M, and Λ:

$$\rho \sim a^{-3(1+\omega)}, \quad \omega = 1/3, \quad \gamma, = 0, \quad M, = -1, \Lambda$$

$$a(t) = t^{2/3}, \quad M, t^{1/2}, \gamma, e^{Ht}, \Lambda$$

(3.8)

$$H(t) = (2/3t), \quad M, H(t) = (1/2t), \quad \gamma, constant\ H, \Lambda$$

The energy density scales as $1/a^3$ for matter and $1/a^4$ for radiation as is physically reasonable. In both cases, the expansion slows the acceleration as the sources become less dense. At large enough values of $a(t)$, matter will become the dominant force over radiation since it dilutes more slowly. Both matter and radiation lead to H scaling as $1/t$ and to power law behavior for a as a function of t. The DE discussion is deferred for now and the focus is on the HBB epochs. A flat Universe has a critical density and the three values for Ω are constants.

The following code plots the density and temperature for radiation and matter alone using current cosmological data. The CMB provides the defined present photon density because the CMB dominates the number density of all photons in the Universe. The present CMB has cooled off during the expansion and now has a temperature of 2.726 degrees Kelvin with a number density, n_γ of about 4.11×10^8 photons/m^3 with an energy density of $\rho_\gamma = 2.6 \times 10^{-4}$ GeV/m^3. The CMB photons have a black-body spectrum of energies and a temperature which increases as time decreases.

The mass density is taken to be both normal "baryonic" matter and DM, which is required by data on the rotation curves of galaxies and by observation of astronomical mass lensing. They require a matter source in excess of what is directly measured by counting up stars and estimating their mass given their luminosities as determined using solar models such as was applied to the Sun in Section 2. Relativistic energy is called radiation here, but the primordial neutrinos mentioned in the nucleosynthesis discussion should be taken into account. They are three doublets of weakly interacting fermions and are known to have masses, albeit small. Therefore, they have no simple equation of state. In this text, a factor is added to the photons which would cover the case of relativistic neutrinos which would add to the "radiation". This factor is 1.68 and it may be imposed by the user in order to see the effect, which is an extreme case. Dark energy is ignored because it has significant effects only at very late times, as will be seen later.

The time when the "matter" and "radiation" densities are equal, t_{eq}, is illustrated in Fig. 3.2, and is about 24,000 years since the HBB, which is taken to be $t = 0$, where $a_o = 0$. An approximate value is $\alpha_{eq} = \Omega_\gamma / \Omega_M \sim 1.47 \times 10^{-4} \sim (t_{eq}/t_o)^{2/3}$, $t_{eq} \sim 24,600$ yr. At t_{eq}, the radiation temperature was about 18,000 K^o, with an energy ~ 1.5 eV. There is an assumed abrupt shift in the behavior of the expansion when the densities are equal. Later work will use MATLAB utilities to solve for $H(t)$ driven by multiple energy sources in order to make the true smoother behavior evident. At $t \sim a$ few seconds, the radiation temperature is on a scale of nuclear binding energies, which means that nuclear physics will be needed to extrapolate backward to earlier times. The previous discussion of nucleosynthesis shows that the HBB history can be successfully explained for times of a few minutes and energies of a few MeV, the nuclear binding energy scale. The SMPP should be able to extend that range to TeV scales using known physics.

```
% find flat space quantities in matter and radiation dominated eras
% ignore dark energy, DE, effects, important at late times
% present Ho in km/sec*million lyr (~30);
% h (~0.73), to ~ 13.8 Gyr;
% assume omega = 1 (flat space)
% To = CBR temp at present
% Ho = 30 .*h , LH = c/H, tH = 1/H, t = n/H, vH = c(1+q), q = (1-n)/
n,
% present time cto
n = 2.0 ./3.0 ; % matter dominated, flat GR, a dependence on t, a/ao
=(t/to)^n
H = 30 .*0.68;
h = 0.68; % present Hubble constant
To = 2.73; % CBR temp in deg K
% Hubble Time in byr ;
tH = 300 ./H   % Hubble Time in byr
```

```
tH = 14.7059
```

```
% q = (1.0 - n) ./n; % Deceleration parameter
% to = tH ./(1.0 + q); % Age in byr
% Temperature at teq - degree K ;
Teq
```

```
Teq = 1.8555e+04
```

```
% Time at Equal Density (no v no DE) - sec ;
teqs
```

```
teqs = 8.2562e+11
```

```
% extrapolate back in rad dominated era, matter n = 2/3, rad n = 1/2
```

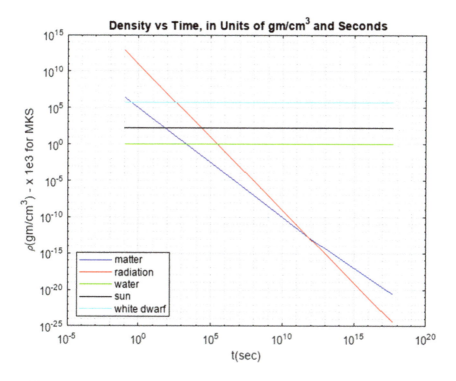

Figure 3.2: Density of the Universe as a function of time for RD and MD epochs.

% for kg/m^3 units simply multiply the y axis by 1e3

As seen in the plot, Fig. 3.3 there is a simple RD scaling — that $T\sqrt{t} = 10^{10}$ in degree and second units. In exploring elevated temperatures, the CMB is opaque about 380,000 years after the HBB because it is in thermal equilibrium due to Thomson scattering of the photons off the electrons in the plasma. However, using nuclear physics and the composition of the Sun, the SMC can be validated from the present to the first few seconds after the HBB. Going from nuclear scales, $\sim$MeV, to atomic scales, $\sim$eV, the HBB model can explore the formation of neutral hydrogen analogous to the formation of light nuclei using Boltzmann equations.

In the case of the temperature, or energy, scaling as the inverse of $a(t)$ occurs because the fluid cools upon expansion, $T \sim 1/a(t)$. This is true for

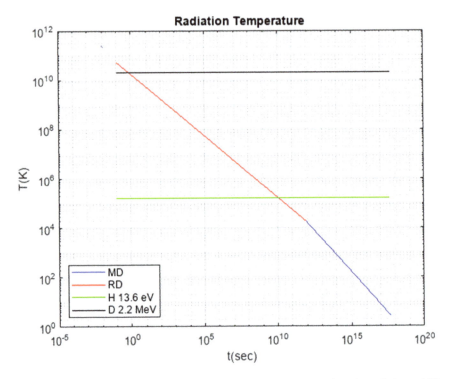

Figure 3.3: Temperature of the radiation of the Universe as a function of time: *MD* and *RD*.

photons, but since the kinetic energy of a non-relativistic system goes as momentum squared, then $T(t)a^2(t)$ is a constant in this case. The momentum of massive particles with respect to the metric decreases during expansion. Similarly, the physical wavelength, Λ, scales as $a(t)$ since waves are stretched or red shifted by the expansion.

The present particle horizon is approximately the physical Hubble distance $D_H = c/H_o$ which corresponds to that physical distance D, where the velocity is c. The time rate of change of D_H is $1/n$, so that the expansion velocity is greater than c. The Hubble distance expands faster than the galaxies since $1/H \sim t$ is greater than $a(t) \sim t^n$ for n less than 1 so that more of the Universe is included inside the horizon and is visible as time goes on. Eventually, the entire Universe becomes visible. This behavior is true for a decelerating Universe composed of matter and radiation. This is not a physical velocity, violating SR, but the velocity of expansion of space itself.

3.3 Hubble History: Hot

The GR geodesic equations for the time evolution of the scale factor $a(t)$ have already been quoted. They are defined by the energy content of the space. Using an equation of state to relate density and pressure, the equation for H can be easily solved for $\alpha(t) = a(t)/a_o = a(t)/a(t_o)$ in the case of domination by a single source of energy. For a power law, $a(t) \sim t^n$, $H_o t = (1 - n)$ is constant. As noted in Fig. 3.3, there is a scaling of $T\sqrt{t}$ in a RD epoch. For a MD epoch, the scaling is as $Tt^{4/3}$:

$$\alpha_m = (3\sqrt{\Omega_m}H_o t)^{2/3}$$

$$\alpha_\gamma = (2\sqrt{\Omega_r}H_o t)^{1/2} \qquad (3.9)$$

$$\alpha_\Lambda = \exp(\sqrt{\Omega_\Lambda}H_o t)$$

In the full general case, with all possible sources, no analytic solution is possible, but MATLAB can easily generate a fine grained solution numerically. In general, for the flat SMC,

$$\frac{d}{dt}a = H_o\sqrt{\Omega_m/a + \Omega_\gamma/a^2 + \Omega_\Lambda a^2}$$

$$(H/H_o)^2 = \Omega_m/a^3 + \Omega_\gamma/a^4 + \Omega_\Lambda \qquad (3.10)$$

For the case of only matter and radiation as sources for the R–W metric, an analytic solution for $H_o t$ is possible rather than the individual power-law behavior shown previously. In the script, the user can choose the numerical matter contribution and observe the change in the plot. Note the smooth transition of $H_o t$ from $1/2$ for RD to $2/3$ for MD. This transition is not abrupt but occurs over about three orders of magnitude in $\alpha(t) = a(t)/a_o$. More matter shifts the transition toward earlier times, as expected. Note the in the RD epoch only photons are accounted for. Relativistic neutrinos can be added, three generations of massless neutrinos, in the script as desired. They will change t_{eq}. Since the masses of the neutrinos are not yet fully determined, this is a worst case estimate.

The solution is found symbolically, output by the script, plotted in Fig. 3.4, and also appears as Eq. (3.11)

$$H_o t = (1/3\Omega_m^2)[(2(\sqrt{\Omega_\gamma + \Omega_m\alpha}) - \sqrt{\Omega_\gamma})^2(\sqrt{\Omega_\gamma + \Omega_m\alpha}) + 2\sqrt{\Omega_\gamma})]$$

$$(3.11)$$

```
% Look at flat space with a mixture of matter and radiation energy
% omegm = 0.315, omg = 4.65e-5 x 1.68 (CMB+3 geneations of v);
syms Om Og c x y z a b
y = int(x/sqrt(Om*x+Og),x); % y = Hot, a = omegm, b = omegg
% definite integral fron alpha = 0 to alpha
% Symbolic implicit Integral for Hot, x = alpha, a = omega m, b =
omega gam
simplify(yy)
```

$$\text{ans} = \frac{2(\sqrt{Og + Om\,x} - \sqrt{Og})^2(\sqrt{Og + Om\,x} + 2\sqrt{Og})}{3Om^2}$$

```
Om = 0.315; % SMC value is ~0.315;
Og = 4.65e-5 ; %.* 1.68; % CMB radiation + 3 v generations
```

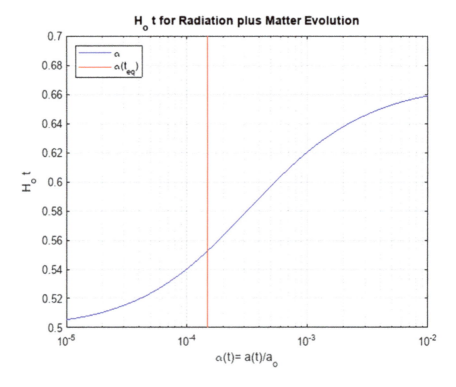

Figure 3.4: Hot as a function of the R–W scale factor in an *RD* to *MD* transition.

3.4 SMC — Matter, Radiation and Dark Energy

In the second Section it was argued that understanding the structure of the Sun required a hot big bang (HBB) model of matter and radiation. Recently, however, the observed expansion of the Universe has forced the addition of a "dark energy", DE. This energy appears to act as a simple repulsive cosmological constant or vacuum energy density. For the present, this simplest of assumptions is made, since it agrees with all the present data. The full equation for H can be integrated numerically in the presence of matter, radiation and dark energy. MATLAB utilities can be used to generate a fine grained full history of $\alpha(t)$ very rapidly. No closed form solution exists, but MATLAB makes it easy to explore the full history of the HBB and a modest extrapolation into the future. Note that the "matter" consists of both ordinary and dark matter, both of which are non-relativistic by assumption. Closed form explicit solutions for $\alpha(t)$ will be used when possible, but implicit solutions $H_9 t = H_o t(\alpha)$ will be used for the full SMC. Using MATLAB makes the numerical evaluation simple so that the full history of the Universe is easily explored.

$$H = \left(\frac{\mathrm{d}}{\mathrm{dt}} a \right) \Big/ a, \quad H = H_o \sqrt{\Omega_m/\alpha^3 + \Omega_\gamma/\alpha^4 + \Omega_\Lambda} \tag{3.12}$$

At early times, the CMB data indicates that photons dominate, thus the hot Big Bang (HBB). However radiation dilutes more rapidly than matter, so that later the Universe is matter dominated, MD. However, the matter is largely a "dark matter", (DM), whose nature, beyond the fact that it gravitates and does not have electromagnetic or strong interactions, is unknown to the SMPP. Finally at late times the cosmological term or DE dominates since, being space itself, it does not dilute under expansion. Again, the nature of this "dark energy" is unknown to the SMPP. However a fundamental particle which has the quantum numbers of the vacuum, the Higgs particle has recently been discovered as discussed in Section 2.

In the script below, the history of the HBB model is integrated out numerically. The user picks the number of sampling points. Time can be extended beyond the present in order to predict the future evolution of the Universe. Additional plots can be made over a wide range of variables as desired. The variables addressed in these plots will all be defined in what follows. Indeed there is only a single variable in the R-W metric, $a(t)$. The plethora of variables can all be derived from $a(t)$, the H_o numerical value, and the present parameter values of the 3 constituents, radiation, matter, and dark energy. Their utility resides in specific applications and they will be defined when needed. A few basic plots follow. The scale factor is shown

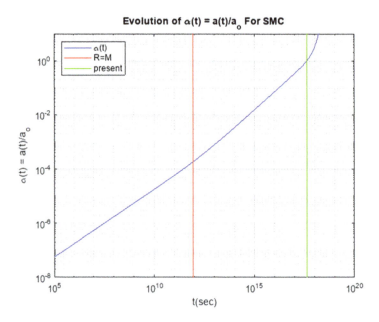

Figure 3.5: R–W scale factor $\alpha(t)$ as a function of coordinate time t.

as a function of coordinate time t in Fig. 3.5, while the conformal time is shown as a function of t in Fig. 3.6.

Conformal time defines an horizon. This definition of conformal time accounts for the expansion of the Universe during the travel time of the photon and restores the 'light cones" familiar from the Minkowski flat space of SR with the limitation that $\tau > 0$ since there is a beginning to time. Light starting from coordinate distance $r = 0$ at $\tau = 0$ arrives at $r = \tau$. Any more distant point is outside the light "cone" and is not yet visible. The conformal clock time slows with respect to the coordinate clock time, t, as the Universe expands. Conformal time. τ is the comoving variable which defines the light horizon. Light travels on null geodesics, as it does in SR, $d^2s = 0 = (cdt)^2 = (adr)^2$, $d\tau = cdt/a = dr$. The conformal time is the comoving radial coordinate for light travel. There is a causality issue that is solved by invoking inflation, as discussed later. The conformal time interval is defined by a definite integral of $cdt/a(t)$. The limits for τ can be 0 and a_o for the interval from the HBB origin to the present. Other variants encountered in the text are 0 and $a(t)$, for the origin to a later time. Using the redshift z, limits of z and 1 form light emission to reception now and define $\tau(z)$.

Alternatively, H can be used to convert $d\tau = cdt/a$ into $d\tau = cda(a^2H)$. For a power law behavior, $H = n/t$, $\tau = ct_o/a_o(1 - n) \sim 1/(1 - n)$ since

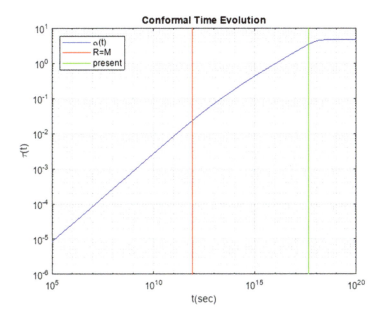

Figure 3.6: Conformal time as a function of coordinate time.

fortuitously $ct_o/a_o \sim 0.95$, or $\tau \sim 3$ for MD with $n = 2/3$ and 2 for RD with $n = 1/2$. The event horizon is $\tau_{\mathrm{eh}} = \int_{t_o}^{\infty} c\,dt/a(t)$. The horizon is the largest comoving distance now visible. The event horizon, as in SR, is the largest distance that will ever be visible. For RD or MD the event horizon is infinite and all is ultimately observable. In contrast, the ΛD horizon is finite, $\sim e^{-Ht}/H$ and the Universe will not ever become fully observable.

```
% used ode45 for more exact cosmology - matter + radiation + dark
energy
% constants
    c = 3.0 .*10 .^8;      % m/sec
    hbar  = 0.66 .*(10 .^-24);  % GeV*sec
    hbarc = 2.0 .*(10 .^-16);   % Gev*m
    to = 4.36 .*10 .^17;   % sec
    ao = c .*to; % m - approx present scale factor
    rc = 5.6;      % critical density GeV/m^3
    rm = 5.6 .*0.315;  % GeV/m^3 - present density of matter
    rr = (4.64 .*10 .^-5) .*5.6;    % density of radiation
    rv = 0.685 .*5.6;              % density of dark energy
    To = 2.726; % CMB temp in deg K
    Eo = 2.35 .*10 .^-13; % convert temp to energy (GeV)
    yr = 3.154 .* 10^7 ; % sec/yr
```

```
% eq setup
  Mp = 1.2 .*10 .^19;      % GeV - Planck mass
  dd = 1.225 .*10 .^-18;   % constant for dA/dt , units /sec
  bb = rr ./rm;            % factors for radiation and vacuum
evolution
  cc = rv ./rm;
tmax = 20; % Maximum Energy Exponent in sec =17.64, present
tmin =  1.0 ; % Minimum Energy Exponent in sec (E~MeV @ 1 sec)
npts = 3000;
tspan = logspace(tmin,tmax,npts); % 1 sec to ~ present, user choice,
 fine grained
[t,y] = ode45(@hubble_m_r_v,tspan,10 .^-10); % start with a small @
tmin
% find rad = matter and rad = hydrogen ionize also decoupling
[yrm,irm] = min(abs(y - 1.47e-4));
[yion,ion] = min(abs(Ee - 1.36 .*10 .^-8));
[ydec,idec] = min(abs(Ee - 0.26 .*10 .^-9));
tion = t(ion) ./yr
```

```
tion = 309.8382
```

```
% Time When Radiation Energy = H2 Ionization (yr) ;
% Conformal time When Radiation Energy = 0.26 eV = decoupling (yr)
tau(idec)
```

```
ans = 0.0755
```

```
% Conformal Time When Radiation Energy = 0.26 eV = decoupling
tau(kk)
```

```
ans = 4.5346
```

```
% Conformal Time at tmax ~ to
tau(irm)
```

```
ans = 0.0187
```

```
% Conformal Time at teq
y(irm)
```

```
ans = 1.4649e-04
```

```
% Scale Factor at teq
```

The rate of change of α for MD is greater than that for RD. Note the rapid increase in the scale after the present due to the DE constituent. The conformal time, τ is related to the coordinate time, t, as plotted below. Note that it levels off in the future under the influence of DE.

Note the flattening of the conformal time in the future due to the effect of DE. There is also an issue about the flatness of the metric and the rapid increase in the curvature with time. The plot below shows that a cosmological constant, when it becomes dominant will drive down a curvature. That effect is more obvious if the user chooses a maximum t value > the present as is the default here. This is a crucial effect of inflation as discussed later. If curvature, κ, exists, the Ω parameters are not constants but are time dependent. For a MD or RD Universe, the curvature increases rapidly. Deviations from flatness are $|\Omega(t) - 1|$.

$$H^2/H_o^2 = \rho/\rho_c - \kappa/a^2, \quad \Omega(t) = 1 + \kappa/[H(t)a(t)]^2 \qquad (3.13)$$

The term in $\Omega(t) - 1$ is plotted below in Fig. 3.7 for the SMC taking $\kappa = 1$. Note that the curvature is drastically reduced by the DE at late

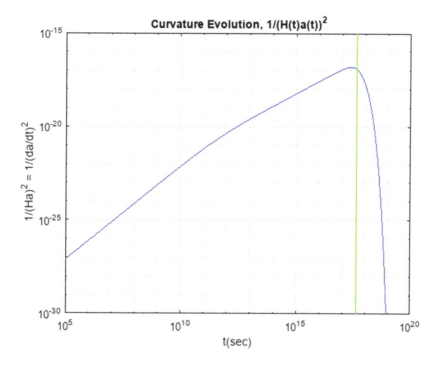

Figure 3.7: Curvature as a function of coordinate time for late time.

times. The vertical line is the present. A dimensionless variable could be plotted as desired, $[c/H(t)a(t)]^2$. In a MD or RD epoch the curvature grows rapidly, making the currently measured flatness of the Universe a puzzle unless early inflation is invoked.

The detailed time dependence of $H_o t$ is plotted below in Fig. 3.8. The value of $1/2$ is expected in a RD epoch as observed. There is a smooth evolution to $2/3$ during a MD epoch. At late and future times the effect of the DE is very evident in in the exponential behavior of the Hubble "constant". This effect makes it possible to observe and infer the presence of DE. Indeed the present epoch is dominated by DE.

Note the vertical line is the present and that the DE induces a late time rapid, $\sim$ exponential, rise. Finally the full evolution of the Universe in terms of Ω in the SMC is plotted at fairly late times. There is a period

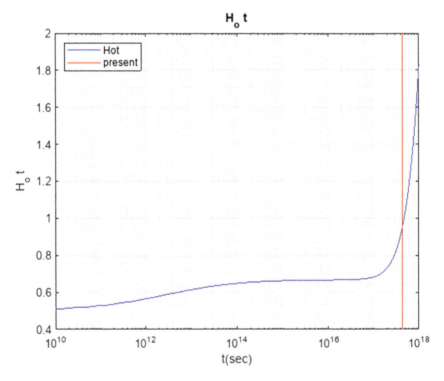

Figure 3.8:   The $H_o t$ product as a function of coordinate time. For RD and MD, values of $1/2$ and $2/3$ are expected.

of RD followed by a MD epoch, with the effect of DE evident only at late times, near the present.

Finally, the Ω parameters for radiation, matter and DE are plotted as a function of t in Fig. 3.9. Note the sequential dominance of radiation, matter, and then DE. At all times, however, the space is flat by construction. Note also that at present the Universe is already in a LD epoch.

```
function dy = hubble_m_r_v(t,y)
global bb cc dd
dy = zeros(1,1);
dy(1) = dd .*sqrt(1.0 ./y(1) + bb ./(y(1) .*y(1)) + cc .*y(1) .*y(1)
);
end
```

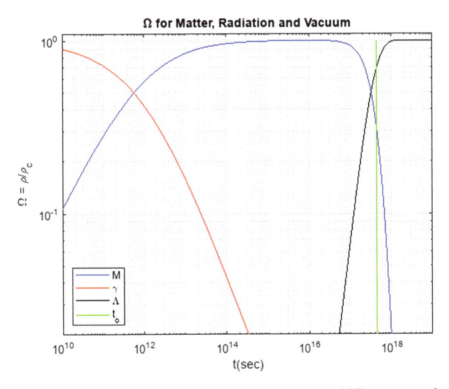

Figure 3.9: $\rho(t)/\rho_c$ as a function of t for the radiation, matter, and DE components of the Universe.

3.5   Age of the Universe

The age of the Universe is finite, assuming that the HBB scenario is correct. The present value of the Hubble constant, H_o, defines a Hubble time $t_H = 1/H_o$. However, this time is derived from the tangent of the curve of $a(t)$ at the present time. Depending on the actual time when H is measured, the value of the Hubble time can be greater than or less than the actual time of the HBB, t_o. Due to the DE and the consequent accelerating expansion, the present result is fortuitous, $H_o t_o \sim 0.95$. If intelligent beings were to make the estimate at an age of $\sim$5 Gyr, they would overestimate by about 2 Gyr. Conversely, if beings mused at 20 Gyr, they would underestimate the age of the Universe. In general, the full dynamic evolution, as given here, is required. The user chooses the time when the inhabitants measure H_o and linearly estimate the age of the Universe. A fine grained numerical integration is again used. An example appears in Fig. 3.10 A full and correct model of the evolution of the Universe is needed to find t_o with examples plotted in Fig. 3.11.

```
% large time range, look at to estimate
% constants
    rm = 5.6 .*0.315;   % GeV/m^3 - present density of matter
    rr = (4.64 .*10 .^-5) .*5.6;   % density of radiation
    rv = 0.685 .*5.6;              % density of dark energy
    yr = 3.154 .* 10^7 ; % sec/yr
    c = 3.0 .*10 .^8;      % m/sec
    to = 4.36 .*10 .^17;   % sec
    ao = c .*to; % m - approx present scale factor
tage = 19.8; % pick age of estimators in byr < 20
[ts,is] = min(abs(t-tage)); % pick nearest point
t(is)
```

ans = 19.7881

```
xint = -b ./slope % intercept
```

xint = 3.2752

The age of the Universe depends on the past history. An accelerating Universe is the oldest, while a decelerating Universe would be the youngest. The SMC posits that there were two phases of acceleration. Initially, there was "inflation" as is discussed later. At present, the DE is causing a period of acceleration. There are also possible closed Universes which collapse, flat Universes, and accelerating Universes. Note that the initial time where the

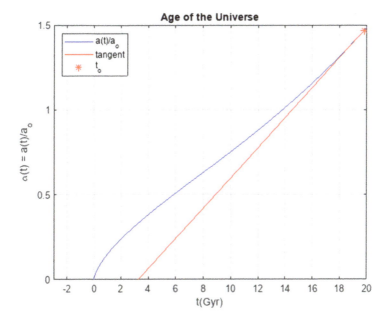

Figure 3.10: Scaled parameter $\alpha(t)$ as a function of coordinate time.

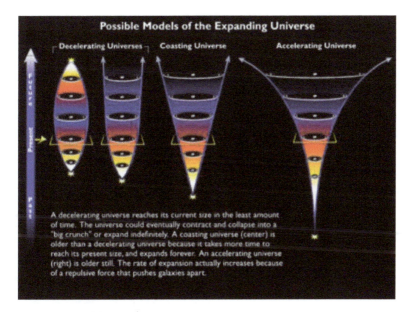

Figure 3.11: Cartoons of the possible ages of the Universe depending on whether it is decelerating or accelerating.

size is zero is small for the closed case and increases steadily up to the accelerating case. Those initial times are the correct ones, and the present state is the same in the four cases. A full and correct model of the evolution of the Universe is needed to find t_o.

3.6   Hot and Curvature

In the SMC, late times are dominated by matter and DE. A possible curvature effect can also be searched for and limits on curvature can be placed. The user chooses Ω for matter and DE. Radiation is ignored for these late times. The curvature arises because $\Omega_M + \Omega_A$ need not equal 1. The current SMC fits to all the available data put limits on the existence of a possible curvature term. From the present, the effects of curvature are most pronounced at larger z values, or earlier times. The script uses the variable z which is often chosen in looking at light emission and reception and is explored in detailed later. The variable z is called the redshift since it is directly the ratio of the wavelength of light at observation, t_o, divided by the wavelength at emission, at time t_e, $1+z = 1/\alpha = \Lambda_o/\Lambda_e$, $\alpha = 1/(1+z)$. But there is only 1 R–W variable and it is simple to transform back and forth. The expression plotted assumes $\kappa = 1$:

$$H^2 = \rho/\rho_c - \kappa/a^2, \quad \Omega(t) = 1 + \kappa/[H(t)a(t)]^2 \tag{3.14}$$

Using light as a probe, looking at earlier emission times enhances the curvature effect because the light travel time is longer, which increases the effect of the expansion on the arrival time. The user can choose several values of Ω_Λ and Ω_M and clear the overlaid curves by using "Run" again as desired. By itself, curvature has a solution $\alpha(t) = H_o t \sqrt{\kappa(1 - \Omega_o)}$:

$$d(H_o t)/d\alpha = 1 \left/ \sqrt{\Omega_m/\alpha + \Omega_\gamma/\alpha^2 + \Omega_\Lambda \alpha^2 + \kappa(1 - \Omega_o)} \right.$$

$d(H_o t)/dz$

$$= -1 \left/ \left[(1+z)\sqrt{\Omega_m(1+z)^3 + \Omega_\gamma(1+z)^4 + \Omega_\Lambda + \kappa(1 - \Omega_o)(1+z)^2} \right] \right.$$

$$\tag{3.15}$$

Plots of both $\alpha(t)$ and z can be made in order to display the apparently different behavior, with a user chosen example in Fig. 3.12. The effects of curvature become most evident to present observers when observing light from earlier times.

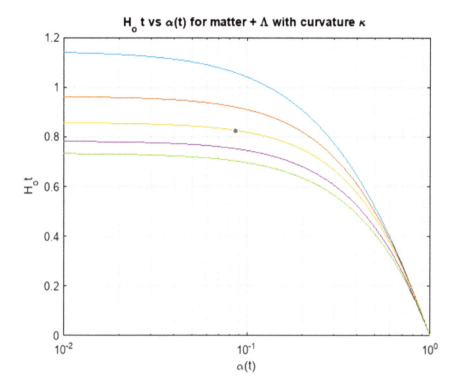

Figure 3.12: $H_o t$ as a function of α or z for several choices of Ω_Λ and Ω_m.

```
% look at later times = matter and dark energy
oml = 0.06;
omm = 0.28;
% find Hot for Universe with curvature
% implicit solution as a function
%  ignore radiation density, omk = 1 - omm - oml
omk = 1.0 - omm - oml  % overdense/underdense
```

```
omk = 0.6600
```

```
% do the numerical evaluation
% lookback time is integral from z = 0 (to) to z = ze (te)
% plot z or alpha
z_alpha = 1; % 0 for z plot, 1 for alpha
```

The user can choose a semilogx plot of either z or α. The effects of a curvature parameter are quite evident for $z > 10$, where $\sim 20\%$ effects on $H_o t$ are observed. For that, limits on the curvature can be set from, for example, supernova data.

3.7   Collapse and Curvature

There is an explicit solution in the simplest case of a closed Universe with curvature $\kappa = 1$ and matter alone using conformal time, τ as the variable. The scale factor and the coordinate time as a function of τ are set by the parameter D, with dimensions of length, which defines the maximum value attained by the scale factor and also sets the collapse time, $D = \Omega_m a_o^3 H_o^2 / 2c^2 = \Omega_m (a_o/2)(a_o/D_H)^2$:

$$a(\tau) = D(1 - \cos(\tau))$$
$$ct(\tau) = D(\tau - \sin(\tau)) \tag{3.16}$$

As seen in Fig. 3.13, the coordinate time rises smoothly to a value of 2π for ct/D, while the R–W factor rises to a value of 2 but then collapses to a point symmetrically in τ. Generally, curvature will be ignored in this text because inflation is later explicitly invoked to solve the problem that the Universe appears to be quite flat while the curvature itself should grow rapidly in an RD or MD epoch. Parenthetically, there is no reason why the experimental parameter Λ might not be negative, which would be attractive and drive the late time Universe toward collapse. It is simply a parameter describing something to which has been assigned the name "DE", but its composition is unknown. Indeed, Einstein introduced a "cosmological term" which was attractive in order to yield a static Universe but retracted it when it was found to be unstable.

```
% closed, k = 1, MD solution, conformal time tau
% D = OmM*ao^3*Ho^2/2*c^2
a = D*(1-cos(tau));
t = D*(tau-sin(tau));
D = 1; % length scale
ylabel('a(\tau)/D')
grid
```

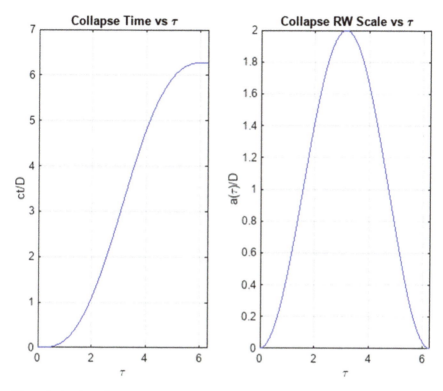

Figure 3.13: Closed geometry and curvature. The collapse occurs in a finite conformal time (left) with a scale that rises to a maximum and then collapses (right).

3.8 Conformal Distance and z

The SMC has a large DE component. The evidence for that comes from observing supernovae as standard candles. The effects of DE are small for z less than ~ 0.1 since the R–W space is locally quite flat. The evidence for DE arises at larger redshifts or large distances. In Section 3.6, the z variable was introduced and the effect of curvature on $H_o t$ was examined for late times, $z > 10$ or $\alpha < 0.1$. Now, one can approximate the situation as a flat space epoch with only matter and DE, which is solvable in closed form (to be done later in Section 4.4). At these values of z, the contributions due to radiation are quite small and can be neglected. The script does the integration numerically, with results shown in Fig. 3.14:

$$H_o dt = d\alpha / \sqrt{(1 - \Omega_\Lambda)/\alpha + \Omega_\Lambda \alpha^2}$$

(3.17)

The symbolic implicit solutions for $H_o t$ which are reported in Section 4.4 are shown in the following, both for the case of an arbitrary mix of matter and dark energy and for the special case of a flat space with $\Omega_\Lambda + \Omega_M = 1$:

$$H_o t = \log \left(\text{OL} + \frac{\text{OM}}{2} + \sqrt{\text{OL}(\text{OL} + \text{OM})} \right)$$

$$- \log \left(\frac{\text{OM}}{2} + \text{OL al}^3 + \sqrt{\text{OL al}^3 (\text{OL al}^3 + \text{OM})} \right)$$

$$= \log \left(\frac{\text{OL}}{2} + \sqrt{\text{OL}} + \frac{1}{2} \right)$$

$$- \log \left(\sqrt{\text{OL al}^3 (\text{OL al}^3 - \text{OL} + 1)} - \frac{\text{OL}}{2} + \text{OL al}^3 + \frac{1}{2} \right) \quad (3.18)$$

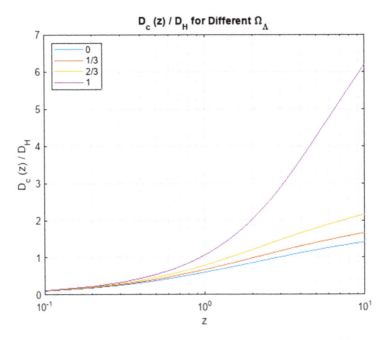

Figure 3.14:   Conformal light distance as a function of z for a few choices of Ω_Λ as shown in the legend.

For the special case of t_o, $\alpha = 1$, the value for $H_o t_o$ is given as follows, which is numerically ~ 0.95 for the SMC:

$$H_o t_o = 1/3\sqrt{\Omega_\Lambda}\,\ln((1 + 2\sqrt{\Omega_\Lambda} + \Omega_\Lambda)/(1 - \Omega_\Lambda)) \qquad (3.19)$$

There are several distance scales in use, but there is only one scale $a(t)$ and a constant H_o. The Hubble distance is $D_H = c/H_o$. A comoving distance, D_c, can be defined and is a function of the z at light emission, with reception at $z = 0$. This distance is not truly commoving nor is it dimensionless. Scaling to D_H would make it dimensionless:

$$D_c(z) = (c/H_o)\int_0^z dz/\sqrt{\Omega_m/a^3 + \Omega_\Lambda} \qquad (3.20)$$

Another distance that is in use is the dimensionless distance for light emission and reception, $\chi = \int_{t_e}^{t_o} c\,dt/a(t)$. In a flat R–W Universe, $\chi = r$, and there is no difference between χ and the actual comoving radial distance r. With a finite curvature, this equality is no longer true. Since the Universe appears to be flat, in the future the comoving distance will be assumed to be r itself.

```
% look at case of matter + DE, radiation small at late times
al = 1.0 ./(1.0 + z);    % scale factor with present = 1, a(t)/a(0)
% find conformal time/distance for light from  z or alpha
```

3.9 Luminosity Distance and SMC

In the literature, there are many variables that may be encountered, a, f, α, τ, z, for example. Indeed, there are several more distance scales in common use in astronomy and cosmology. The luminosity distance, D_L, refers to the observed luminosity compared to the actual, intrinsic, luminosity which is understood for stars using solar models as in Section 2 or used with supernovae as distance scales of known intrinsic luminosity, or "standard candles".

The luminosity distance is evaluated for the SMC with the possible addition of curvature. Symbolic solutions are shown for matter, DE, and curvature separately as was done for power-law behavior. Then the user-chosen mix of these three sources for the luminosity distance as a function of z is displayed. The many distance scales, it should be remembered, all flow from a single function: the R–W scale factor $a(t)$. In addition, at small z values, near the present time, the cosmological effects are small and the different distances collapse into a single common shape with classical distance $= ct$ behavior. The implicit expression for $H(z)$ with radiation, matter, DE, and curvature is:

$$a(z) = a(z)/a_o = 1/(1+z), \quad D_H = c/H_o$$
$$H(z) = H_o\sqrt{\Omega_\gamma(1+z)^4 + \Omega_m(1+z)^3 + \Omega_A + \Omega_\kappa(1+z)^2} \tag{3.21}$$

The luminosity distances for pure matter, D_{Lm}, $D_{L\Lambda}$, or curvature, $D_{L\kappa}$, are functions of the z parameter as is $\alpha(z)$ and, normalized to D_H are:

$$D_{Lm}(z) = (1+z)\int_0^z (1+z)^{3/2}, \quad D_{L\Lambda}(z) = (1+z)z,$$
$$D_{L\kappa}(z) = (1+z)\int_0^z 1/(1+z) \tag{3.22}$$

The angular diameter distance for an object of proper length L_{pr} subtending an angle θ on the sky is $L_{pr}\theta$ which scales as $a(t)$ at emission due to Hubble expansion. The angular diameter distance, D_A, is then reduced by a factor $(1+z)$ from the conformal time, $a_o\tau(z)$. The luminosity distance, D_L, for an object of known luminosity, L (a standard candle), and observed flux F, such as a type I supernova, is defined as it would be for a flat space, $D_L^2 = L/(4\pi F), D_L \sim a_o\tau(z)(1+z)$. The flux suffers two factors of $(1+z)$: one is due to the redshift energy decrease of the photons and the second is due to the Hubble expanded spread in time of the photon flux. Those two flux reduction factors mean that the luminosity distance is enhanced by a factor $(1+z)$.

The complete luminosity distance, excluding radiation but including matter, DE, and curvature, is defined by the integral

$$D_L(z) = (1+z)\int_0^c 1/\sqrt{Q}, \quad Q = \Omega_m(1+z)^3 + \Omega_\Lambda + (1 - \Omega_\kappa(1+z)^2) \tag{3.23}$$

The comoving distance, $D_c(z)$, the angular distance, $D_A(z)$, the luminosity distance, $D_L(z)$, and the look-back distance, $D_{LB}(z)$, are variants in common use and are defined as follows. The look-back distance depends on both the emission and reception, or present time for light:

$$D_c(z) = D_H\int_0^z (H_o/H(z))dz, \quad D_A(z) = D_c(z)/(1+z),$$
$$D_L(z) = D_c(z)(1+z), \quad D_{LB}(z) = D_H\int_0^z dz/(1+z)^2 \tag{3.24}$$

First, the distances for three individual contributors, matter, DE, and curvature, are found symbolically using "int". The user then enters the values for matter, DE, and curvature, and the luminosity distance is plotted. The MATLAB numerical utility "integral" is used for that plot since analytic solutions are not possible. Radiation is ignored at these relatively late times, but it could be included in the numerical evaluations. The angular and luminosity distances for small z are $\tau(z) \sim (c/H_o)(z/a_o)$, $D_A \sim a_o\tau(z)/(1+z)$, $D_L \sim a_o\tau(z)(1+z)$, both of which are $\sim a_o\tau(z)$ for nearby small z values.

```
% look at luminosity distance (measurable)
% for difference M and DE fractions, curvature
% first symbolics, one at a time as with the Ho power laws
syms z x
dlm = (1+z) .*int((1+x)^-1.5,0,z)
```

$$\texttt{dlm} = -(z+1)\left(\frac{2}{\sqrt{z+1}} - 2\right)$$

```
dll = (1+z)*z;
dlk = (1+z) .*int((1+x)^-1,0,z)
```

$$\texttt{dlk} = \log(z+1)(z+1)$$

```
omm = 0.3; % Enter Omega Matter
oml = 0.7; % Enter Omega Dark Energy
omk = 0; % Enter Omega (1-Omegao), curvature
Hz = @(z) 1 ./sqrt(omm .*(1+z) .^3 + oml + (1-omk) .*(1+z) .^2);
for i =1:length(z)
    dl(i) = (1+z(i)) .*integral(Hz,0,z(i));
end
```

The different distance scales are then computed in a purely MD geometry. In that case, $H(z) = H_o\sqrt{\Omega_m}(1+z)^{3/2}$, and the conformal transit time is $\tau(z) \sim (1 - 1/\sqrt{1+z})$. The z range is now expanded for the plot to the full coverage of the Webb telescope, $z \sim 20$. Indication of why it is widely used. Plots of the variables as a function of z appear in Fig. 3.15.

```
tauz = 1 - 1 ./sqrt(1+z);
tauang = tauz ./(1+z);
taulum = tauz .*(1+z);
```

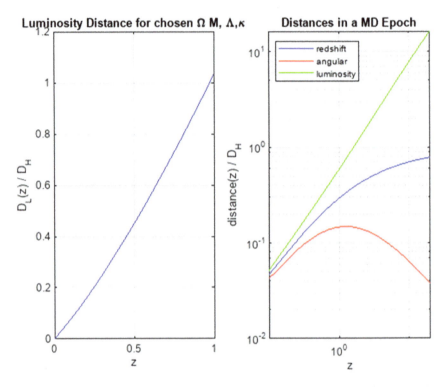

Figure 3.15:   Luminosity distance as a function of z, for the chosen values of matter, DE, and curvature (left). Distance scales as a function of z (emission) in an MD geometry (right).

3.10   Conformal Time

Several distances are used in the SMC; the redshift distance, the angular size distance, and the luminosity distance were defined and plotted previously in Section 3.9 in terms of the z or redshift variable. Alternatively, the conformal time can be used as the variable, which is useful in exploring issues of causality. The R–W metric is $d^2s = -(cdt)^2 + a^2(t)[d^2r + r^2 d^2\Omega]$. The variable a has dimensions of length and r is a dimensionless comoving radius. Physical distance is $a(t)dr$. Light travels on a null geodesic with $d^2s = 0$ or $a(t)dr = cdt$. The differential conformal time, τ, is defined to be $cdt/a(t) = dr = d\tau$ and the conformal time interval between light emission and present reception is $\int_{a_e}^{a_o} cd't/a(t') = \tau_o - \tau_e$. During the light travel time, the R–W metric has evolved, which the conformal time accounts for. In terms of H and z, $cdt/a(t) = cda/Ha^2$ and $dz = -da/\alpha^2$. For matter alone, and assuming a flat metric:

$$\tau = \int cdt/a = c \int dz/H(z) = \int cda/Ha^2$$

$$(3.25)$$

$$H(z) = H_o\sqrt{\Omega_m(1+z)^3}$$

This result for conformal time is called tauz, $\tau(z)$, or conformal time as a function of emission at redshift, z, approximately $\tau(z) \sim (c/H_oa_o)z$. This conformal time definition here is specific and has light emission at z and reception at $z = 0$. Other definitions can be used to define the definite integral. The present $\tau(z)$ is defined as a definite integral where in the past the limits of the integral were implicitly $t = 0$ and $t = t_o$ and were not shown for τ since the whole history of the Universe in the HBB was assumed. Using symbolic integration, the integral is displayed. For power-law behavior, $\tau(z) = \int_a^{a_o} cdt/a(t) = \tau[1 - (t_z/t_o)^{1-n}] = \tau[1 - (1+z)^{-1/\beta}]$, $\beta = n/(1-n)$. For matter alone, the solution is $\tau_{MD}(z) = (2c/H_o\sqrt{\Omega_m})(1 - 1/\sqrt{1+z})$, or $\tau_{MD}(z) \sim (2c/H_o\sqrt{\Omega_m})z$ at small z. The angular size measure is $D_A(z) = \tau(z)/(1+z)$ and the luminosity measure is $D_L(z) = \tau(z)(1+z)$. For small values of z, $D_c(z) \sim z$ for pure MD, while for pure LD, the dependence is always linear in z. All these different measurements must vanish at $z = 0$ and increase with increasing z at small z. As telescopes like the Webb explore larger z values, the sensitivity to the properties of DE will increase.

```
% look at Dc(z)/D_H, populate z from ~ first stars to present
% MD behavior of H(z), tau = int(dz/H(z)) = int(cdt/a) = int(cda/Ha
^2)
% tauz = 2(1 - 1 ./sqrt(1+z));
```

The conformal distance is the light distance from $z = 0$ and o to z. The angular distance and the luminosity distance at a location z are simply related to the conformal distance. The light travel or look-back distance is the distance from emission at a source, z, to reception now, $z = 0$. The script finds conformal distance $D_c(z)$ symbolically as a function of z for a flat metric with a user-chosen Ω_m. Results are plotted in Fig. 3.16.

```
% find conformal time/ distance
Hz = Ho*sqrt(OM*(1+z)^3);
DCM= int(1/Hz,0,z)
```

$$\text{DCM} = -\frac{2\left(\frac{1}{\sqrt{z+1}} - 1\right)}{\text{Ho}\sqrt{\text{OM}}}$$

```
Hz = Ho*sqrt(OL);
DCL= z/(Ho*sqrt(OL))
```

$$\text{DCL} = \frac{z}{H_o\sqrt{\text{OL}}}$$

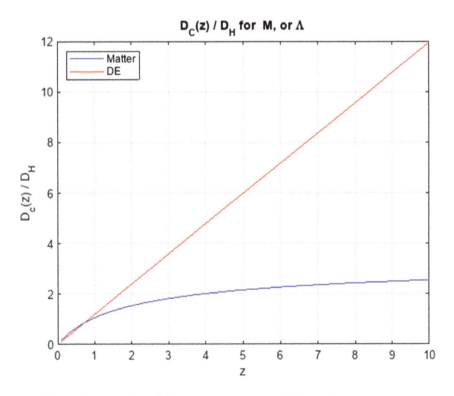

Figure 3.16:   Conformal distance as a function of z for purely matter or DE.

3.11 Light Emission and Reception in SMC

Light propagation in a purely matter-dominated space is formulated as a function of z which has an analytic solution. The z variable is typically one of the choices for looking at light propagation. Conformal time has also been used, as shown in the previous section. The present range of precise optical observations is approximately $z \sim 2$, but new telescopes like the Webb will continue to dramatically improve the reach in z. Already at that z, the light received is from a distance $\sim$80% of the Hubble distance, and the emission occurred at about 20% of the present time or $\sim$2.8 Gyr.

The recently deployed Webb telescope, shown in Fig. 3.17, will improve this reach and view the earliest star and galaxy formation. This telescope operates in the infrared because the redshift of these early objects is too large for normal optical operation. The Webb orbits the Sun at the stable L2 Lagrange point. It is 1.5 million kilometers further away than the orbit of the Earth and sits stably in the shadow of the Earth. This location allows the optics to operate at $-233\ C^0$ which enables IR operation. Recently, the Webb discovered a black hole which was already formed at $t \sim 0.5$ Gyr. At $z \sim 20$, the time would be only $\sim$180 Myr from the HBB.

Figure 3.17: The Webb telescope, built of many individual IR reflectors.

This script specializes to an MD epoch only, since an analytic solution exists. A more complex solution for a flat space with both matter and DE also exists but is not used just now. At long look-back time the effects of DE on the evolution of the Universe are small, but the present time is actually in an DE-dominated epoch. In the MD case, the scale is simple, $a(t)/a_o = a(t) = (3H_o t/2)^{2/3}$. Since $1 + z = 1/\alpha$, the emission time is $t_e(z) = (2/3H_o)(1/(1+z)^{3/2})$ and the "look-back" time is simply $t_{LB}(z) = t_o - t_e(z)$. At small values of z, the look back time approaches zero as z/H_o:

$$t_{LB}(z) = (2/3H_o)(1 - 1/(1+z)^{3/2}) \qquad (3.26)$$

Numerical corrections due to radiation and DE should be made for accurate interpretation of the data. The plots, shown in Fig. 3.18, are now made for the full expected z range of the recently deployed Webb IR telescope.

```
% light emission and reception in an expanding MD cosmology
% DH = c/Ho , Hubble distance = 4439 Mpc
% n = 2 ./3;   a/ao = (t/to)^n,  MD only power law
% tH = 1/Ho = Hubble time = 14.52 Gyr
zlb = 1./((1+z).^(3/2));
to = 0.6666.*tH;
te = to*zlb;
tlb = to -te; % lookback time
```

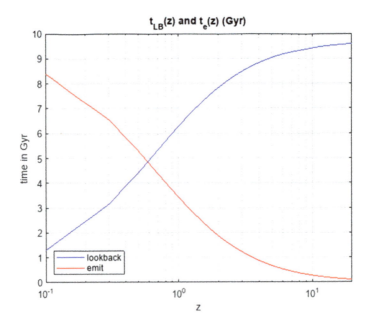

Figure 3.18: Emission and look-back times as a function of z at emission.

At large z the lookback time rises toward t_o, in this simplified treatment $\sim$9.5 Gyr, which approaches the entire history of the Universe. At z of zero, reception is instantaneous following emission and the look-back time is zero. The low z look-back time is approximately linear in z, $t_{LB}(z) \sim (c/H_o)z$. The look-back time approaches the present time at large z, as expected, scaling as $\sim(2/3H_o)(1 - 1/z^{3/2})$.

3.12 Horizons in SMC in Matter-Dominated Epoch

The plot made is a cartoon showing light traveling from an emission point, r_e, at a time t_e in the past. During the travel time, the space expands. The light may or not reach the observer at the present time. Event horizons may exist, but true horizons do not exist in an MD metric. In general, there are no horizons if the space is flat and if one waits a sufficient time. An event may not yet arrive at a specific observer time, but it will later, although the space expands during the light travel time making the coordinate (not comoving) distance traveled longer. The user chooses the emission time and the emission distance. The movie cartoon follows the light from emission time to the present, with the last frame shown in Fig. 3.19.

Later, the discussion of inflation shows that with an exponential increase of the scale $a(t)$, a true horizon exists, and in some sense, the Universe expands during inflation faster than the speed of light. Another example of a space with true horizons is mentioned in Section 4.6 where a static de Sitter space is explored.

```
% look at light rravel in a matter dominated
% movie for in and out of horizon distance w.r.t. present time;
te = 0.5 % Emission Time / Present time , te/to < 1
```

```
te = 0.5000
```

```
% find emission distance to arrive at present time = 1
tau = 3; % MD conformal time, comoving horizon, power law, n = 2/3
taue = tau .*(te .^0.333) % emission tau, for emiission time te
```

```
taue = 2.3817
```

```
tauz = tau - taue % location re to just arrive at r = 0 at to
```

```
tauz = 0.6183
```

```
% comoving distance From emission to receive now = tauz
re = 0.7 % emission comoving distance from observer at r = 0
```

```
re = 0.7000
```

```
% time runs from emission time  to present or reception time
t = linspace(te,1.0); % time from emission to present
r = re - tau .*(t .^0.3333 - te .^0.3333); % photon path
% photon may or may not reach r = 0 by to
DHt = t ./0.666; % Hubble distance at t
% now plot Hubble distance and the photon
```

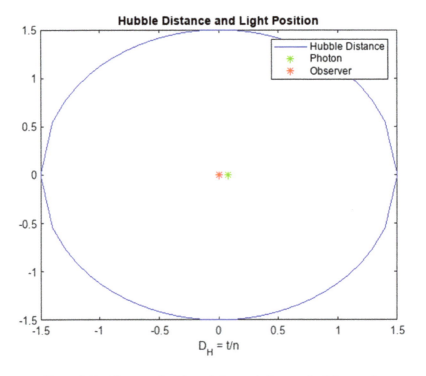

Figure 3.19:   Cartoon of conformal time and distance for light travel.

To conclude this section, the basic R–W metric has been explored using the parameters of the HBB, SMC. The many variables that might be encountered in the literature were defined and explored. These variables are those relevant to large-scale aspects of the Universe, time, scale factor, conformal time, conformal distance, light travel time, look-back time, redshift, and several others. Next, the evidence supporting the SMC is explored in Section 4, and the CMB structures which determine many of the parameters of the SMC are examined in some detail in Section 5. The modifications arising when considering inflation and the induced quantum and gravitational fluctuations have been deferred until Sections 6–8.

Chapter 4

History and Future with Dark Energy

"Equipped with his five senses, man explores the universe around him and calls the adventure Science."

— Edwin Powell Hubble

"Astronomers can look back in time. We can look at things as they used to be. We have an idea there was a Big Bang explosion 13.7 billion years ago. We have a story of how galaxies and stars were made. It's an amazing story."

— John C. Mather

4.1 SMC Parameter Measurement

There have been a lot of definitions and formulae up to now. Much of it is due to the plethora of variables used and the units adopted, with $G = 1$ for example. It is past time to turn to the SMC, first the HBB and then the current "inflation" model. First, some comments on the sources in GR that drive the metric. The SMC is defined by the observational data and is created to explain it. The SMC contains DM, for which no SMPP candidate exists and no search has found evidence for, as with the Wimp and SUSY null searches, discussed later in Section 5. Nevertheless, DM exists, as inferred from galactic rotation curves and gravitational lensing. The DM is dark; it is defined to have no strong or electromagnetic interactions. Whether or not it has weak interactions is unknown at present, since its existence has only been inferred by its gravitational effects. Most of the experimental DM searches assume that it participates in weak interactions.

The rotation curves indicate a DM "halo" extending well beyond the radii of the visible matter (see Fig. 4.1). This DM is needed to control the

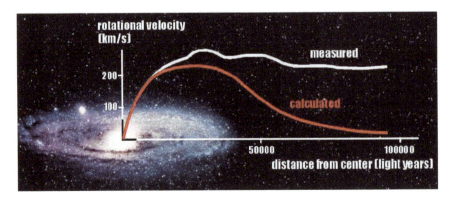

Figure 4.1: Rotational velocity of a galaxy as a function of distance from the galactic center.

centrifugal force of the rotational velocity as inferred from the Doppler-shifted spectra of the stars in the galaxy. Dark matter is needed basically to hold the rotating galaxies together.

In surveys of galaxies, much evidence for lensing of background sources by foreground, but invisible, DM exists. An example is shown in Fig. 4.2. The light deflection angle of a light ray with impact parameter b by a point-like body of mass M is $\theta_M = 4M\hbar c/(M_p^2 bc^2)$, as shown in Section 2.1. That relationship defines the focal length of the DM. For a distance D between the source and the observer with an intervening mass M at a distance xD from the observer, if M is on the line of sight of the observer, a ring of angular size $\theta_E = \sqrt{(4G/\mathrm{Mc}^2)((1-x)/\mathrm{Dx})}$ is predicted: the Einstein ring. If the intervening mass is not on the line of sight, then multiple arcs of the image are expected and observed.

For Dark Energy, DE, there is compelling evidence from studies of super-novae as "standard candles". Using these as probes, there is strong evidence of a repulsive acceleration. Since there is no SMPP candidate for DE, the simplest assumption is made. There is a repulsive cosmological constant Λ which is not a dynamical object at all, but part of the metric of the Universe, $H_\Lambda = \sqrt{\Lambda/3}$. There is, at present, no evidence for an equation of state for DE. The data are, at present, consistent with $\omega = 0$. Until the data compel a revision, the simplest assumption should be made and confronted with more precise data as it becomes available. The model for DE implies a constant value for H when DE dominates at late times.

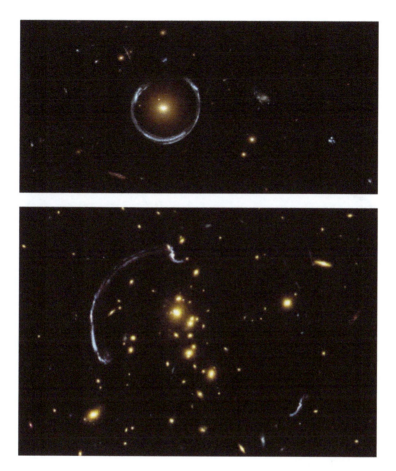

Figure 4.2: A gravitationally lensed "Einstein ring" of light from a galaxy caused by intervening DM (top) and arcs of deflection for off-axis observation (bottom) — Hubble Space Telescope.

The data on supernovae as a function of z displayed in Fig. 4.3 show that there is an accelerated repulsive effect at late times. The present data go out to $z \sim 2$, but the Webb telescope will greatly extend that range by $\sim 10\times$. The literature prefers to use the variable z, which was previously introduced for that reason. Plots with different values of Λ have already been shown, for example, in Section 3.7.

The strongest evidence for DE occurred in studying the z dependence of type I supernovae and led to a recent Nobel prize for the definitive discovery of DE. The supernovas act as "standard candles" of known luminosity

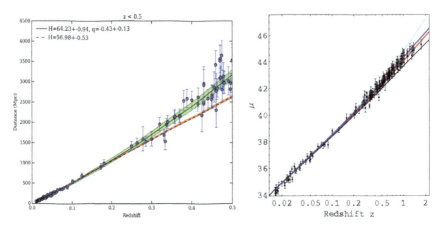

Figure 4.3: Distance of supernovae as a function of z. The red line on the left, lower z data, and the black line on the right, higher z data, are for a model with no DE and show that the data require a repulsive DE component.

and the luminosity distance can then be used to determine the distance. Parenthetically in an expanding Universe, time intervals increase since $dt/a(t) = dr$, a constant. This increase has also been observed in SN brightness curves which are intrinsic. The conformal time z dependence of matter in an MD phase has a z dependence different from that for DE. This is a good reason why z is a heavily used variable in cosmology, although the default time variable is t or τ in this text. The conformal z time is $\tau_z^M = (2c/a_oH_o)\sqrt{\Omega_M}(1 - 1/\sqrt{1+z})$, while for DE, $\tau_z^\Lambda = (c/a_oH_o)z\sqrt{1 - \Omega_M}$, both having the same small z limiting behavior, scaling as $(c/a_oH_o)z$.

The DE may evolve with an equation of state. Present limits on that evolution appear below. The data are not yet sufficient to distinguish among several models. A more precise measurement of the DE energy density as a function of z is needed. The result of these data-driven measurements is the present SMC. The combination of supernovae data, CMB data, and the results of galaxy survey data greatly restrict the SMC parameters. The curvature is also, for simplicity, assumed to be zero, and the Universe is flat within the present measurement error. The conclusion that the Universe is flat has largely been assumed in this text, although some calculations with a non-zero κ value were made in Section 3. The experimental limits on an equation of state for the DE are presently consistent with a simple cosmological constant which does not evolve, as shown in Fig. 4.4, where the equation of state parameter is given constant term, ω_o

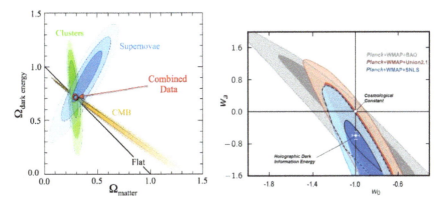

Figure 4.4: SMC derived from the combined data (left) and DE equation of state, $\omega = \omega_o + \omega_a z$ (right).

and a slope in z. The data are seen to be consistent with a simple constant value. This simplification will be assumed in this text. However, the data presently do not make strong restrictions on the DE equation of state. The plot is for an assumed linear expansion of the DE, horizontal, constant term, vertical, slope with z.

The CMB is sensitive to the matter content as are the galactic clustering data. Combined with the supernova data, the parameters of the SMC are constrained to be flat. The Friedman equations in a flat space, $\kappa = 0$, relate the geometry, the scale factor $a(t)$, to the sources, matter density, pressure and Λ. They were quoted in Section 3 and are repeated here for reference. They describe the expansion, the acceleration of the scale, and the continuity of the density and pressure:

$$H^2 = \left[\left(\frac{d}{dt}a \right) \Big/ a \right]^2 = (8\pi/3M_p^2)\rho + \Lambda c^2/3$$

$$\left(\frac{d^2}{dt^2}a \right) \Big/ a = -(4\pi/3M_p^2)(\rho + 3p/c^2) + \Lambda c^2/3 \qquad (4.1)$$

$$\frac{d}{dt}\rho = -3H(\rho + p/c^2)$$

The SMC has been used to describe the evolution of the Universe. However, there are unexplained features that are simply assumed. Why is the Universe flat? Why is the CMB so very uniform over all space when the whole sky contains regions that could never have been in causal contact in the HBB model? These questions can be answered by postulating "infla-tion", a period of rapid growth of the scale factor, much like that driven by the DE exponential growth. This hypothesis will be examined in later sections.

4.2 Evidence for Dark Matter

DM is inferred from galactic rotation curves and lensing of light from distant sources, as mentioned in Section 4.1. The plots in Figs. 4.5 and 4.6 are extremely schematic, but the basic conclusions are clear. The velocity of a galaxy, which extends beyond the visible stars in the galaxy, implies material which gravitates but does not radiate: DM. The distorted image of a foreground galaxy implies deflection of the light by intervening matter, matter which is not itself visible, DM.

```
%  evidence for dark matter , galaxy rotation curves
% Dark Matter is About 1/4 of the Universe by Mass
% Dark Matter was First Inferred from Galactic Rotation Curves
% Dark Matter i Also Inferred from Lensing,  Einstein Ring Sizes
% Dark Matter is Searched for, to no avail yet, in Scattering
Experiments, e.g. WIMPs
```

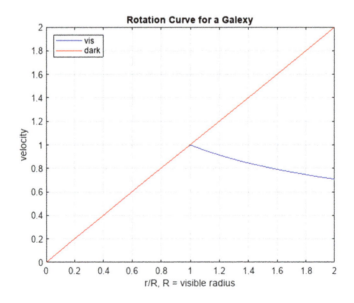

Figure 4.5: Schematic of a rotation curve. The matter velocity should fall off outside the mass distribution, not so for a DM "halo".

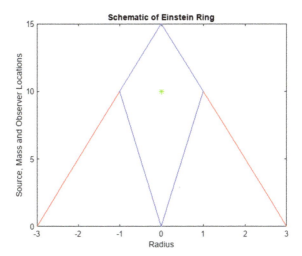

Figure 4.6: Schematic of an Einstein ring. Light from a source is bent by an intervening source (green star for DM) with a focus whose location depends on the mass of the DM.

There is DM which gravitates, but the SMPP has no such candidate. What is it? In regard to WIMP searches, it is normally assumed that the rotational velocity of the Milky Way galaxy, ~300 km/s, is with respect to a DM at rest. This then imparts an incident kinetic energy of the DM WIMP with respect to the detectors. After scattering off the nuclei in the detector, the recoil atoms in the detector have an energy of ~0.1 MeV, which is a small signal and necessitates locating the detectors far underground in order to reduce the cosmic ray background and the neutrino flux. So far, in more than a decade of increasingly sensitive searches for DM particles using nuclear target recoils, as seen in Section 5.1, we have not seen a significant signal for particles with masses greater than ~10 GeV. Lower mass than that imparts a detector recoil energy typically too small to appear above experimental backgrounds. Recently, an august review body has advocated pushing the search limits down to, and perhaps below, the neutrino "floor".

```
% The Milky Way Galaxy Has a radius of ~ 30,000 LY
% The Galaxy Rotates every 200 Million Years
ly = 9.46 .*10 .^19;  %  m
T = 6.3 .*10 .^19; % period, sec
vdm = (2.0 .*pi .* 30000.0 .*ly) ./T;   % galactic velocity m/sec
vdm = vdm ./1000  % km/sec
```

vdm = 283.0425

```
% Velocity of Galaxy w.r.t. Dark Matter (km/sec)
% kinetic energy of a 200 GeV DM particle
Tdm = (200000 .*(vdm ./(3 .*10 .^5)) .^2) ./2.0
```

Tdm = 0.0890

```
% Maximum Kinetic Energy of 200 GeV DM Particle (MeV)
```

Searches for DM have expanded to cover a lower mass range since heavy DM searches have not found a detectable signal. For example, experiments at the CERN LHC have active searches for light DM particles, particles which the recoil experiments are less sensitive to. An example is shown in Fig. 4.7. The search was for a decay mode of the Higgs boson into a photon and a "dark photon". The dark photon interacts only weakly and is therefore invisible to the detector. The limit set is for dark photon masses less than $\sim$10 GeV, with a Higgs decay branching ratio of less than $\sim$3%.

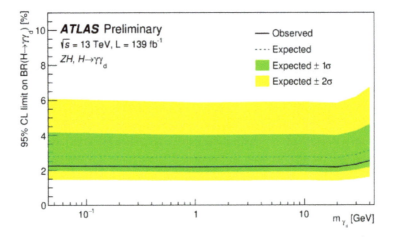

Figure 4.7:   Upper limits on the branching ratio for a Higgs boson to decay into a photon and a dark photon as a function of the dark photon mass.

Many other searches at other accelerators or dedicated single-purpose experiments are in progress or being planned to search for light DM particles. An example of expected dark photon limits is shown in Fig. 4.8.

Other proposals aim to use r.f. techniques to search for even lower masses, with axions as a possible DM candidate. Without a specific SMPP

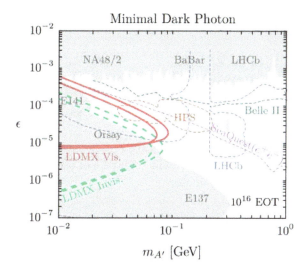

Figure 4.8: Sensitivity to a dark photon above a 10 MeV mass at a variety of existing or proposed experiments.

candidate, a "full court press" is called for. Unless the DM possesses weak interactions, these experiments will not detect DM, and some other techniques need to be invoked.

4.3 Variable Conversion Tool

There are many variables in use and conversion between them is a useful utility for the reader to have access to. In this script, the user can choose from a list of possible variables. The desired numerical value is also then chosen, and the corresponding values for all the other variables are computed and displayed. The general evolution for the SMC with matter, radiation and DE is evaluated for all eight cosmological variables, $z, \alpha, t, T, H, D_H, K, \tau$ as a function of time. Using MATLAB, one can simply evaluate the equations of motion numerically over the full history of the Universe. It is hoped that this is a useful utility for the reader when confronted with a large variety of cosmological variables. The new variable here is K which is the wave number of a perturbation and is used in the later discussions of inflation. Note that neutrinos are not treated properly, partially because the masses of the three generations of neutrinos are unknown. That being the case, the photon energy density can be multiplied by a factor to account for the extreme case of all massless neutrinos. The user can then see the approximately maximum shifts due to the neutrino uncertainties.

```
% Add 3 Generations of Neutrinos
%    rr = rr .*1.68;
```

The user can restore that scaling and see if variables of interest are changed significantly.

Some results at present, at RD/MD equality and at photon–baryon decoupling are given in Table 4.1, which may be of use to the reader to set the scale of some of the parameters.

Table 4.1: SMC parameter values at present, decoupling and R/M equality.

Quantity	Present — Power Law, RD, MD	Decoupling	Equality
t/t_o	1	2.93×10^{-5}	1.31×10^{-6}
τ	3.35	0.076	0.019
α	1	9.1×10^{-4}	1.47×10^{-4}
z	0	1100	6778
$1/H$ (Mpc)	4439	0.2	0.010
$K = \alpha H$ (Mpc^{-1})	0.00022	0.0045	0.0147
Horizon exit	1/4439		
T(eV)	2.34×10^{-4}	0.259	1.60
ρ(GeV/m^3)	5.57	7.45×10^9	1.77×10^{12}

```
% find conversion factors for cosmological variables
% z, a, t, T, H , k ,DH, or tau
% constants
    mp = 0.938; % nucleon mass, GeV
    hbarc = 2.0e-16 ;   % GeV*m
    c = 3e8; % m/sec
    hbar = hbarc ./c;
    Mp = 1.22e19;  % Planck mass, GeV
    Mpc = 3.086e22 ;  % Mpc in m
    to = 4.36e17;   % to in sec, Hoto = 0.95, to = 13.81 Gyr
    ao = c .*to;
    To = 2.35e-4;   % temp in eV
    DHo = 4439;   % present horizon in Mpc
    rho = 5.6 ;   % present density in GeV/m^3
    rm = rho .*0.315;        % GeV/m^3 - present density of
matter
    rr = 4.64 .*10 .^-5 .*rho;   % density of radiation ( factor
1.68 if 3 v)
```

```
% Variable to Evaluate','z','t/to', 'a/ao','rho(GeV/m^3)','T(eV)',
'D_H(Mpc)','k(Mpc-1)','tau'
itype = 1; % intput variable, find all other corresponding
% first do the numerical evaluation
% solve Hubble eq: d(alf)/d(t), mic of DM, DE, matter, rad
tmax = 17.64;
tmin = 0;
z(imin)  % z
```

ans = 3.9883

```
tt(imin) % t/to Value
```

ans = 0.1116

```
yy(imin) % a/ao Value
```

ans = 0.2005

```
rh(imin) % Density Value (GeV/m^3)
```

ans = 695.1066

```
T(imin) % Temp Value (eV)
```

ans = 0.0012

```
DH(imin) % DH Value (Mpc)
```

ans = 705.8777

```
K(imin) % K Value (Mpc-1)
```

ans = 2.8400e-04

4.4 Solutions for DE + Matter

The SMC postulates a mix of matter and DE which dominate at late times as required by the supernovae data. The dynamics for the scale factor $a(t)$ can be solved in a closed form implicitly as $H_o t$. The results were quoted in Section 3.8 without proof and a simple numerical integration was done there. Now, the analytic solution is found by using a change of variable from α to $y = \Omega_M + \alpha^3 \Omega_\Lambda$ which is a strategy that enables MATLAB to find a symbolic solution. Making the transformation, a definite integral is found with upper and lower limits of integration:

$$H_o(t_o - t) = (\sqrt{\alpha}) \int d\alpha / \sqrt{\Omega_M + \Omega_\Lambda \alpha^3}$$

$$= (1/3\sqrt{\Omega_\Lambda}) \int_{yl}^{yu} dy / \sqrt{y(y - \Omega_M)},$$

$$yl = \Omega_M + \Omega_\Lambda \alpha^3, \quad yu = \Omega_M + \Omega_\Lambda \tag{4.2}$$

The following result shows that the DE makes a comparable contribution to the scale factor as matter does as early as $\alpha \sim 0.5$, $z \sim 3$. Indeed, this fact enabled the successful supernova discovery of near present-day accelerated expansion of the Universe. In turn, this led to the postulate of DE. Indeed, we are in a ΛD epoch at present:

```
% matter + DE - numerical
% variable change y = OM + OL*alpha^3 needed for MATLAB
% Implicit definite Integral for Hot, function of OL, OM, and alpha
imHot = iu-il
```

$$\text{imHot} = \log\left(\text{OL} + \frac{\text{OM}}{2} + \sqrt{\text{OL}\,(\text{OL} + \text{OM})}\right)$$
$$- \log\left(\frac{\text{OM}}{2} + \text{OL al}^3 + \sqrt{\text{OL al}^3\,(\text{OL al}^3 + \text{OM})}\right)$$

```
OM = 1-OL; % add flatness assumption
Hot = eval(imHot)
```

$$\text{Hot} = \log\left(\frac{\text{OL}}{2} + \sqrt{\text{OL}} + \frac{1}{2}\right)$$
$$- \log\left(\sqrt{\text{OL al}^3\,(\text{OL al}^3 - \text{OL} + 1)} - \frac{\text{OL}}{2} + \text{OL al}^3 + \frac{1}{2}\right)$$

```
OM = 0.315;
OL = 0.685;
aleq = (OM ./OL) .^0.3333 % equality of terms on H for MD and LD -
integrand
```

```
aleq = 0.7719
```

`Hot(1)  % M + DE`

`ans = 0.9510`

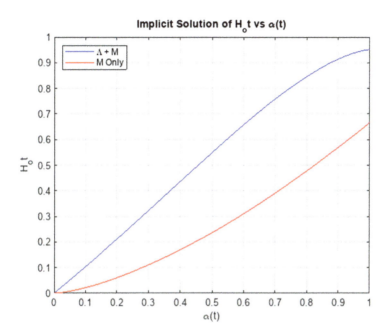

Figure 4.9: Plot of $H_o t$ as a function of $\alpha(t)$ for matter alone and matter plus DE.

The MD result of $2/3$ for $H_o t_o$ is observed at $\alpha(t) = 1$, while with DE, the result is 0.95, as was already seen using strictly numerical techniques. At $\alpha(t) = 0.3$, the total value of $H_o t$ is only $1/3$ due to M and $2/3$ due to DE. Thus, at $z = 2.33$, the effect of DE is already dominant, as seen in Fig. 4.9.

4.5 $H(t)$ for Matter Plus Dark Energy

Dark energy has an effective matter density $\rho_{\mathrm{DE}} = c^4 \Lambda M_p^2/8\pi$. The conformal distances for late times, with matter or DE dominance, are $D_C(z) = c \int_0^z \mathrm{d}z/H(z)$, with $H(z) = H_o\sqrt{\Omega_m(1+z)^5 + \Omega_\Lambda(1+z)^2}$. The conformal distances for pure matter or pure DE are solved symbolically in closed form using the symbolic utility "int". The combined effect is found numerically in this script, although a closed-form solution was already found in Section 4.4:

```
syms z OM OL Hz Ho DC c
Hz = Ho*sqrt(OM*(1+z)^5);
DCM =c*int(1/Hz,0,z)
```

$$DCM = \frac{2c\left(\frac{1}{(z+1)^{3/2}} - 1\right)}{3\,Ho\,\sqrt{OM}}$$

```
Hz = Ho*(1+z)*sqrt(OL);
DCL = c*int(1/Hz,0,z)
```

$$DCL = \frac{c\log(z+1)}{Ho\,\sqrt{OL}}$$

The solution with both M and DE is done numerically, assuming a flat space. The user chooses the value of Ω_Λ. The value of z_{eq} is here defined to be when the two sources inside the square root, $(1 - \Omega_\Lambda)(z+1)^3$ and Ω_Λ, are of the same magnitude:

```
% Flat space with a mixture of matter and vacuum energy
% supernova + CMB wiggles indicate omegav ~ 0.685, omem = 0.315
omegv = 0.68;
% flat space equality of contributions of DE and M to H
zeq = ((1-omegv) ./omegv) .^0.333333
```

```
zeq = 0.7778
```

```
aeq =1/(1+zeq)
```

```
aeq = 0.5625
```

```
% Scale When DE and Matter Contributions are Equal
Ho = 2.23e-18; % Ho in sec^-1
tH = 1.0 ./( Ho .*3.156e16) % Hubble time in Gyr
```

$$tH = 14.2088$$

```
% Hubble Time, in Gyr
% specify the fraction of critical density in the vacuum
% present time t, not n/H due to vacuum energy - explicit solution
```

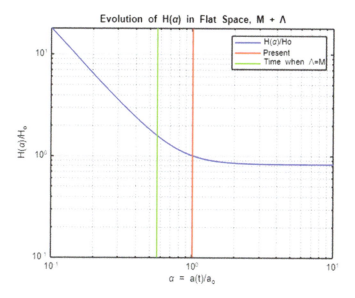

Figure 4.10: Plot of $H(\alpha)$ as a function of α. The effects of matter and Λ are = at $\alpha \sim 0.6$.

The falloff of H with time is $\sim 1/t$ in pure MD, which is reduced with DE and H approaches a constant value when DE dominates (see Fig. 4.10). DE in the SMC has a constant H.

4.6 Horizons in a deSitter Metric

The concept of a horizon is important in both SR (special relativity) and GR, especially in the initial inflation phase of the SMC. Light in all cases is assumed to move on null geodesics. In SR, a radial photon has a path, $d^2s = 0 = (cdt)^2 - d^2r$, so that SR coordinate velocity is always c, $dr/dt = c$. For emitted light to be received at some time depends on the separation distance, which defines the "particle horizon". If there is a definite end to the light cone, there is an "event horizon" since all of space cannot yet be seen. With enough time, there is no horizon in SR.

In the $R-W$ metric, the dimensionless conformal time τ allows for changes in $a(t)$ during the light travel times. For light travel from emission, the photons also travel on null geodesics with $dr/dt = c/a(t)$. The conformal time from $t = 0$ to t is then defined to be the definite integral, $\tau = \int_o^t (cdt)/a(t)$. For a power-law behavior of the scale factor, $a(t)$,

$\tau = ct/a_o(1-n)$. For a constant H, $\tau = (c/H)[1-e^{-Ht}]$. Constant H will later be assumed in the first simplified model of inflation.

In the case of a Schwarzschild metric, there is a curvature spatially, and the null radial geodesic is $(dr/cdt) = (1 - r/R_s)$ with an event horizon, a black hole, at the Schwarzschild radius, R_s. The SMC appears to be flat, with zero curvature where there is no horizon. The conformal time definite integral in terms of a, t, or z at emission, with reception at t_o or $z = 1$ is

$$\Delta r = r_o - r = c \int_t^{t_o} dt/a(t) = (c/a_o) \int_0^z dz/H(z), \quad \tau(z) \sim cz/(H_o a_o)$$

(4.3)

A counter-example is a deSitter space, which has a curvature. It has a positive cosmological constant. The radial metric is $d^2s = (cdt)^2(1 - r^2/a^2) - d^2r/((1 - r^2/a^2))$. This is similar to the Schwarzschild metric in SR and it also contains a horizon. The coordinates for time and space are sinh and cosh of (t/a), respectively. The deSitter acceleration is $\frac{d^2}{dt^2}a = c^2\Lambda a/2$ with solution $a(t)/a_o = e^{c\sqrt{\Lambda/3}t}$ and the scale factor grows exponentially. The resulting horizon is an example of a GR horizon, but only as an academic exercise, not assumed to pertain to the cosmology of the Universe. The null geodesic in this metric is $dr/dct = (1 - (r/a)^3)$ which is zero at r equal to a. Indeed, for some choices of emission location and emission time, the light never arrives in a finite time. In the script, the user chooses a light emission location and time. A movie of the path of the light is then made. Reception time at $r = 0$ is then found. It is possible that the light is never received because the space itself inflates too rapidly. In this case, the movie is not made. The movie when made is in equal time steps indicated by the green stars and their spacing shows that the light is slowed by the rapid expansion during the light transit time.

In a ΛD epoch, H is a constant. Define $a(0) = 1$. In general, the light path is expressed in $1/H$ units. For reception at time to at $r = 0$. For small values of the exponents, $r_e \sim t_o - t_e$. There is an event horizon since as t_o grows, t_e approaches a limit beyond which information is never available, which is the result of the exponential growth of the scale. In a deSitter epoch, space expands so rapidly that there are events which can never catch up to the observer.

The user chooses the emission radius and time, r_e and t_e.

```
% the deSitter metric-has an horizon
% constant H assumed. Define t in 1/H units, a(0) = 1
```

```
% emit at time te, position re
% receive at time tr, position r = 0
re = 0.56;
teh = log(1 ./re) % event horizon
```

t_{eh} = 0.5798

```
% time of emission
te = 0.54;
tr = - log(-exp(-teh)+ exp(-te))
```

t_r = 3.7833

```
% Reception Time at r = 0, = tr)
rrr = re -(exp(-te) - exp(-ttt));
```

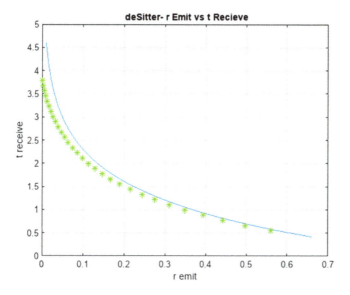

Figure 4.11: Light reception r (comoving) as a function of light emission time.

The blue line represents the horizon in emission location and reception time in order to reach the reception location at $r = 0$. The dots are the "movie" in equal time steps for a photon in the expanding deSitter space (Fig. 4.11). As with all movies, it can be replayed using the controls supplied, in particular the playback speed.

Chapter 5

From the Hot Big Bang

"Unified theories also show us why we observe the World to be
governed by a variety of 'fundamental' forces of apparently
differing strengths: inevitably we must inhabit a
low-temperature world ... and at these low energies the
underlying symmetry of the World is hidden."

— Frank Tipler

"The Universe is under no obligation to make sense to you."

— Neil deGrasse Tyson

In Section 2.5, the HBB model was used in concert with nuclear physics to
predict the primordial helium abundance successfully. In Sections 3 and 4,
the Robertson–Walker (R–W) metric was defined and the parameter values
used in the SMC were explained. Although the identities of dark matter
(DM) and dark energy (DE) remain elusive, the SMC is very successful in
its agreement with cosmological data. In Section 5, some candidate DM
particles are defined and the searches for them are touched upon. The
current understanding of the Universe is crucially dependent on the idea
of "inflation", a model which is used to explain the faint structures in the
photons of the CMB and the similar structure in the large scale structure
(LSS) of the matter of the Universe. This inflation paradigm is of such
importance that Sections 6–9 go toward exploring it.

5.1 Supersymmetry and WIMPs

The SMC has DM as a main source for GR. It has gravitational mass but is
otherwise inert to the strong and electromagnetic force. In the SMPP, only
the weak force remains. WIMPs are weakly interacting massive particles
postulated to exist as the DM particles. For the last decade and more,

particle physicists have been searching for "WIMPs". The idea of a WIMP is related to the fact that the SMPP has no candidate for the DM that has been postulated to account for the observed lensing of objects where no visible matter was present. The rotation curves of galaxies also had velocities which extended far beyond the visible stars of those galaxies. The data indicated the need for gravitational mass that was "dark". Assuming that the DM also had weak interactions, it could be observed in collisions with ordinary matter, via Z^o exchange, for example, by observing the recoils of target nuclei observed in a background-free detector. These detectors were typically located far underground where the only backgrounds were from neutrino scatters off the target nuclei.

How heavy were the WIMPs? The SMPP has three forces with three different coupling constants. In addition, these coupling constants "run" because the basic reaction vertices are surrounded by a sea of virtual quanta. The theory of the SMPP dictates how the couplings "run" with energy. Figure 5.1 shows the results of the renormalization group equation (RGE) calculations. The forces do not simultaneously converge. A possible modification to the SMPP is to add a new symmetry: supersymmetry (SUSY). It is proposed to be active at a mass scale of a few TeV and would have associated particles with a new SUSY conserved quantum number. In addition, the modified RGE also unifies the SMPP coupling constants at a high mass scale, comparable to the Planck mass, assuming, of course, that no new physics appears over the highest ten orders of magnitude in energy. This convergence led to the hope of a grand unified theory (GUT) unifying all the forces of the SMPP. This theoretically motivated model led to a series of experiments searching for WIMPs using nuclear recoils from WIMP scattering. In addition, experiments at the LHC and elsewhere have directly searched for SUSY or DM particles that decay into detectable SMPP particles.

A WIMP that is produced in the Hot big bang (HBB) will stay in equilibrium with the photon/neutrino entropy until its weak interaction rate falls below the Hubble constant, at which point it becomes stable and observable rather like the helium nuclei which were explored previously in nucleosynthesis. In the following code, the user picks the WIMP mass and the weak cross-section for production. The cross-section is a generic weak scale cross-section. The freeze-out temperature is found as well as the relic WIMP number density, which is compared to the measured relic DM density. In this simple model, WIMP masses in the hundreds of GeV range, with cross-sections of order pb ($1b = 10^{-28}\text{m}^2$), will supply the required

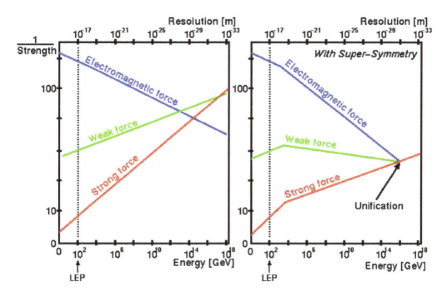

Figure 5.1: Results of RGE for the three SMPP forces. Left — SMPP, Right — addition of SUSY particles at masses ~TeV, which then unify the SMPP forces at a mass scale ~10^{16} GeV.

DM density. The success of the nucleosynthesis study gives confidence in an HBB model for WIMP production, although the mass scale is now TeV and not MeV.

Dimensionally, a typical electroweak cross-section is ~$\alpha_w/(2M_{wi}^2)$, where the electroweak coupling constant is α_w and a threshold to make a pair of WIMPs of mass m_{wi} exists. In general, for processes with a coupling constant α, the cross-section for a massless particle goes as $\alpha^2 T^2$, while for massive particles with a propagator of mass M, the cross-section goes as $\alpha^2 T^2 M^4$ if $M \gg T$. The decay width for a massless particle is, dimensionally, $\Gamma = n\sigma = \alpha^2 T$, while for massive particles, it is $\alpha^2 (T/M)^4 T$. Since $H \sim T^2/M_p$ in RD, the ratio $\Gamma/H \sim \alpha^2 (M_p/T)$ for massless particles and $\alpha^2 (T^3 M_p/M^4)$ for massive particles. These simple generic arguments enable a discussion of the conditions for freeze-out and decoupling generally. For a WIMP mass of 100 GeV, the energy density/entropy ratio ~3.8×10^{-10} GeV for a $1\,pb$ cross-section. A WIMP mass of 100 GeV is therefore a possible DM solution. Clearly, the mass of the WIMP is not well constrained by the requirement to be the DM. There is little experimental guidance for those researchers searching for WIMP detection. However, as temperatures drop to ~0.1 GeV, the Universe consists of about 26% DM, photons, electrons, neutrinos, and NR neutrons and protons. In the early stages of expansion, RD, the WIMP will be relativistic with a density scaling as the fourth power of T and contributing to H as the square of T; $\rho_{\rm rel} = \pi^2 g_* T^4/30$, $H_{\rm rel} = \sqrt{(8\pi^2 g_*/90)}(T^2/M_p)$.

The user picks a WIMP mass and cross-section. After this a full surface of mass and cross-section and the resulting candidate DM density are displayed. Clearly, this exposition is very approximate, although it yields the correct orders of magnitude. WIMP masses of hundreds of GeV and production cross-sections of pb are indicated. The search for WIMP and SUSY interactions at both accelerators and underground detectors extends from mass scales of $\sim$10 GeV to $\sim$1 TeV and is, presently, a null search. WIMP Γ and H are shown in Fig. 5.2 as a function of T, while WIMP densities as a function of mass and cross section appear in Fig. 5.3.

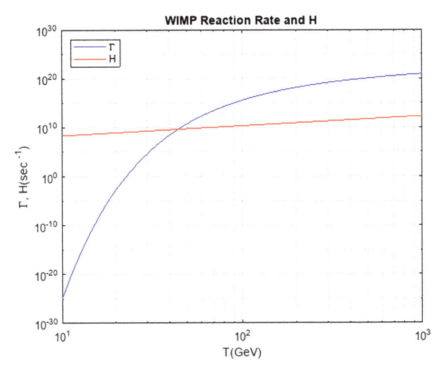

Figure 5.2: WIMP reaction rate and H as a function of T. Equilibrium is lost at an energy of $\sim$45 GeV for a WIMP mass of 1000 GeV.

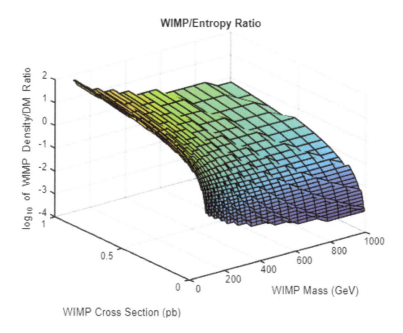

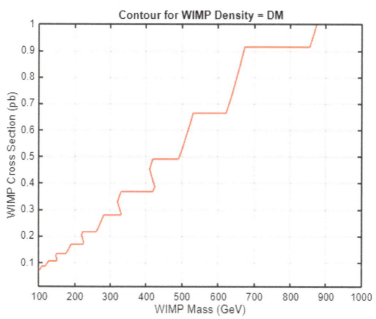

Figure 5.3: Surface of WIMP density/DM density as a function of WIMP mass and cross-section (top). Contour for ratio = 1 (bottom).

```
% Find freeze out temp and abundance of WIMP
% constants
    c = 3.0 .*10 .^8 ; % m/sec
    k = 8.3 .*10 .^-14 ; % 1/GeV
    hbarc = 2 .*10 .^-16 ; % Gev*m
    hbar = 6.6 .*10 .^-25 ; % Gev*sec
    Mp = 1.2 .*10 .^19 ;      % GeV - Planck mass
    % entropy = 2*pi*pi*gst *T^3/45
    gst = 100 ; % relativistic DOF in SM
    so = 2.9 .*10 .^9 ; % present entropy density in m^-3
    rats = (5.6 .*0.265) ./so ; % ratio of DM mass density to
entropy - no baryons
% s*a^3 = constant in adiabatic expansion
% S-B law, rho = pi*pi*gst*T^4/30 - energy density
% H in radiation domination ~ T^2/Mp
Mw = 1000; % Wimp Mass in GeV ;
T = linspace(Mw ./100, Mw,1000); % temp in GeV, NR freeze - M-B
stats
% WIMP number density
% WIMP cross section
% find EW mass of wimp
```

mew = 1.8257e+04

```
% Wimp Mass if EW (GeV)
% find wimp reaction rate - compare to H to see if freeze
% Freeze out Temp (GeV)
nwfr = nw(ifr)
```

nwfr = 1.4315e+43

```
% Freeze out Wimp Number Density /m^3
```

sfr = 4.8922e+53

```
% Freeze out Entropy /m^3
ratsfr = nwfr ./sfr
```

ratsfr = 2.9261e-11

```
% Freeze out Wimp Density/Entropy ;
ratms = ratsfr .*Mw
```

```
ratms = 2.9261e-08
```

```
% Freeze out Wimp Mass Density/Entropy
rats
```

```
rats = 5.1172e-10
```

```
% Present DM Mass Density/Entropy = rats
```

```
ratW_DM = 57.1806
```

```
% WIMP Mass Density / DM Mass Density
% now look at Wimp = DM abundance vs mass and cross section
```

Cross-sections of 0.1–1 pb can provide sufficient DM from WIMPs for WIMP masses from 100 to ~800 GeV. The structure in the curves is just digital noise due to the binning using temperature. As seen in Fig. 5.4, in this mass range, the present cross-section limits are approximately 10^{10} times more stringent than this simple dimensionally argued estimate. If

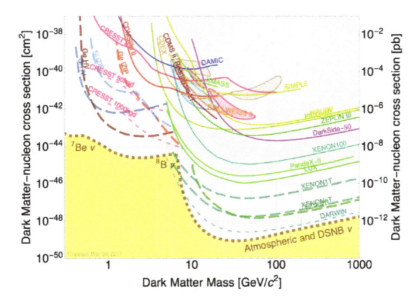

Figure 5.4: WIMP cross-section limits as a function of mass. The neutrino "floor" is indicated.

WIMPs have weak interactions, they are clearly heavily suppressed by some mechanism.

Alas, it was a great idea, a GUT and a new mass scale just now within reach at the LHC, but it was not to be. After many null experiments, shown in Fig. 5.4, the WIMP searches have pushed far below the needed weak interaction cross-section of ~1 pb and have begun to hit a "neutrino floor", set by the irreducible neutrino background. For masses above about 10 GeV, a WIMP DM is excluded. In fact, searches are now evolving to be sensitive to lower masses and other postulated candidates for the DM are being searched for with different techniques. It was a wonderful idea, brought down by a series of experiments. However, a recent scientific panel has put a premium on further efforts to go to or below the "neutrino floor". In parallel, many, many searches for SUSY particles at the LHC and elsewhere have been unrewarded at mass scales up to ~a TeV, with production cross-sections less than a pb. The focus is widening to lower mass WIMPs at accelerators, where the increased sensitivity available at a high intensity accelerator argues for searches at those facilities.

The spectrum of solar neutrinos, Fig. 5.5, forms a background to deep underground WIMP search experiments. The increase in background at low neutrino energy is mirrored in the increased value of the atmospheric neutrino limit at low WIMP mass. In particular, the Be and B neutrinos

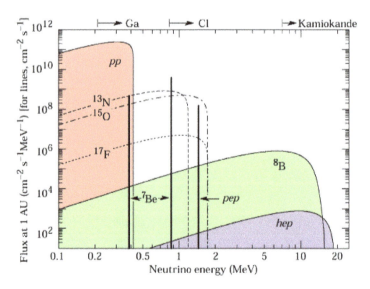

Figure 5.5: Solar neutrino fluxes as a function of energy.

at WIMP masses ~0.5 and 6 MeV are shown in the WIMP low mass limit plot, Fig. 5.4.

Direct production of SUSY particles is excluded up to ~1–2 TeV mass by LHC experiments. Recent LHC data taking has logged integrated luminosities >100 fb^{-1} at an energy of 13 TeV. Typical exclusion limits on SUSY masses are therefore ~1 TeV as shown in Fig. 5.6.

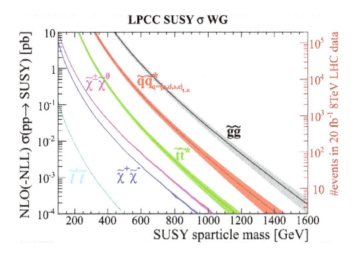

Figure 5.6: Production SUSY cross-sections as a function of SUSY particle mass for different species of SUSY particles.

Since the DM mass is unknown, the search for DM has extended over a full mass range, from the high mass WIMP searches down to a very low mass axion like mass range. There is a "full court press" going forward for DM assuming it possesses weak interactions. A summary limit plot shown below in Fig. 5.7, indicates the breadth of the limits — from MeV down to 10^{-17} eV. At present, there are only upper limits over this impressive mass range.

5.2 The CMB and "Baryons"

Previously, the helium fraction was estimated in the discussion of nucleosynthesis, which occurred at energies of ~1 MeV, less than the helium binding energy due to the large photon entropy. The fluids then continued to evolve in the HBB, as a mix of the dominant photons, with minority protons and helium nuclei along with electrons and weakly interacting neutrinos. A similar situation occurs as further expansion cools the photons

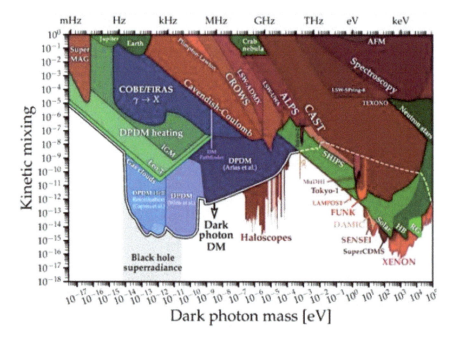

Figure 5.7: Limit plot for dark photon masses as a function of the strength of the DM photon mixing with an SMPP photon.

and finally stable hydrogen with a binding energy of 13.6 eV can form. After nucleosynthesis, the SMPP predicts that p are free or bound into light nuclei. Neutrons have decayed or are bound up in light nuclei. The HBB model then assumes a plasma of protons, electrons, neutrinos, and photons. At energies much less than the hydrogen binding energy, neutral hydrogen forms.

When the plasma scattering rate is $\Gamma_\gamma < H$, the photons fall out of equilibrium and approximately free stream thereafter. This is the same approximation applied to the WIMP searches. The photon fluid is then transparent and the CMB appears at a mean photon energy of ~ 0.26 eV at a time of $\sim 400{,}000$ years since the HBB. The CMB observed today is the earliest time that can be observed using light since the charged plasma was previously optically opaque. If primordial neutrinos could be observed, their weak interaction scale of $\sim$MeV would open a new view to times of $\sim$ minutes as was the case for helium nuclei. With the advent of gravitational wave detection, the possibility of "seeing" much earlier than the CMB decoupling time also becomes possible. However, the large wavelengths involved make the construction of "antennas" rather more difficult than the detection of binary black hole formation shown in Section 2. For now, the optical photons of the CMB are the focus.

In principle, the GR continuity equation could be applied to the electrons, $\frac{d}{dt}n_e + 3Hn_e = -\langle\sigma v\rangle(n_e^2 - n_{eo}^2)$, where the cross-section times velocity uses the Thomson cross-section for NR $\gamma - e$ scattering and the velocity comes from Maxwell–Boltzmann statistics, $\langle\sigma v\rangle \sim 4\pi^2\alpha_{em}B_H/(m_e^2\sqrt{3m_eT})$. The non-equilibrium number density sources the time dependence. However, as with the nucleosynthesis analysis, a simple approximation can be made. The full Boltzmann transport equations are not solved here for the reactions $e + p \longleftrightarrow H + \gamma, \gamma + e \longleftrightarrow \gamma + e$. The relevant cross-section is for Thomson scattering, $\sigma_T = (8\pi/3)\alpha_{em}^2\lambda_e^2$, where α_{em} is the fine structure constant, $\alpha_{em} \sim 1/137$, and λ_e is the electron Compton wavelength. The relevant binding energy is $B_H = 13.6\,\text{eV}$ for atomic hydrogen. The simplification is to assume thermal equilibrium and use the Maxwell–Boltzmann distribution, as was done for nuclei, Section 2.5. The number density ratio of e, p, and H are then

$$[n_H/(n_en_p)](1 - X_e)X_e = (\hbar c)^3[(m_ec^2\text{kT})/2\pi]^{-3/2}e^{B_H/\text{kT}} \tag{5.1}$$

In this equation, X_e is defined to be the fraction of free electrons, $= n_e/(n_e + n_H) = n_p/n_b$. Since $n_p = n_e, n_H = n_p(1 - X_e)/X_e$, the binding energy competes with the photon entropy $\eta = n_b/n_\gamma = n_p/(X_en_\gamma)$. Solving for X_e results in a quadratic equation: the Saha equation. Note that the number density of "baryons", n_b, actually refers to charged SMPP particles which interact with the photons, that is, electrons and protons. Baryons, such as protons and neutrons, are SMPP bound states of three strongly interacting quarks. Nevertheless, this is the convention in use and it is adopted here. The photon number density in equilibrium at temperature T is $n_\gamma = g\zeta(3)T^3/\pi^2$. Solving the quadratic as a function of z, the Saha approximation can be examined. (The Riemann zeta functions, ζ, arises from the integration of the distribution functions and are a MATLAB special function.)

$$1 - X_e = X_e^2 \left[4\sqrt{2/\pi}\zeta(3)(n_b/n_\gamma)(\text{kT}/m_ec^2)^{3/2}e^{B_H/\text{kT}}\right] \tag{5.2}$$

The decoupling of the plasma occurs at $kT \ll B_H$ as expected. The scale factor used in the script is conventionally chosen as z, $1 + z = 1/\alpha(t) = a_o/a(t)$. The CMB decoupling occurred when the Universe was ~ 1100

times smaller than at present, estimated using the Saha technique as 1400. As shown in Fig. 5.8, the decoupling time, z, is a function of the user chosen baryon fraction. The user chooses the Ω_b value in order to see the effect on the decoupling time. The dependence is made explicit in Fig. 5.9. This dependence serves as a crucial input to a consistency check of the present cosmological model. In this approximation, there are no free electrons remaining and they are all taken up in hydrogen. Note that the DM is not part of the calculation since it only gravitates by construction. The Saha approximation is used here as a useful first estimate of decoupling of matter and radiation.

In what follows, decoupling is simply defined to occur at a photon energy of $0.26\,\mathrm{eV}$. This occurs at a $z \sim 1100$ which happens at a time $\sim$400,000 years after the HBB. Much detail is thereby evaded, for example the last scattering occurrence for the CMB.

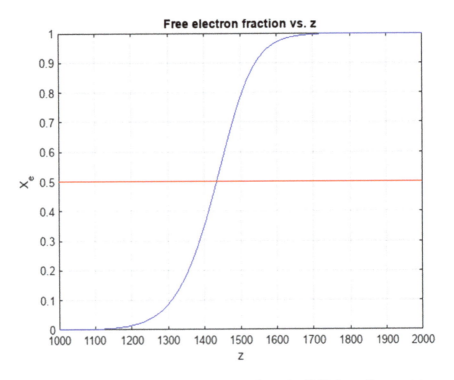

Figure 5.8:   Free electron fraction as a function of z in the CMB decoupling transition.

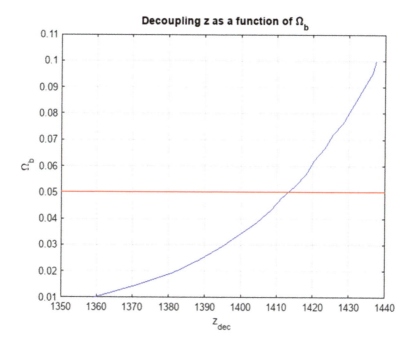

Figure 5.9: The dependence of the Saha estimate of z_{dec} on the value of Ω_b.

```
% look at CMB decoupling in the Saha approximation
% to the complete Boltzmann eq
% constants
    kb = (1.0 ./40) .*(1.0 ./300);  % Boltzmann k in eV/oK
    h = 0.68;  % Hubble factor - 10% spread
    To = 2.75;   % present temp in Kelvin
    B = 13.6 ;  % H binding energy (eV)
    me = 511000.0 ;  % electron mass in eV
    zet = zeta(3);  % Riemann function
omegb = 0.09;  % Present Baryon Density/Critical
Xe = (-q + sqrt(q .*q + 4.0 .*q))./2.0;  % quadratic solution
 % find 50% point, define as decoupling
zdec = zspan(imi)
```

zdec = 1.4343e+03

```
% z at decoupling defined to be Xe = 1/2
% look at decoupling dependence on Omega b
```

The decoupling time using the Saha analysis depends on the "baryon" fraction which is a useful cross-check of the SMC model. In what follows, decoupling is defined to occur at a photon energy of 0.26 eV. This occurs at a $z \sim 1100$ which happens at a time $\sim$400,000 years after the HBB. Much detail is thereby evaded, for example the last scattering occurrence for the CMB.

5.3   Boltzmann Equation for the CMB

Boltzmann transport is needed to go beyond the Saha approximation for the CMB. The simplest such single equation has the electron fraction sourced by a reaction rate Γ with an H "dilution" of H. The equilibrium electron fraction is X_e^o. The CMB photons follow an energy equation in the variable y sourced by the non-equilibrium value of X_e simplified here to be a single Boltzmann transport equation. At the CMB, the decoupling scale, $z \sim$ 1100, was at $a(t) \sim 4.4$ Mpc. The user again chooses Ω_b. Much more on the CMB Boltzmann equations comes later after a discussion of the crucial effects of inflation on the detailed small quantum variations in the structure of the CMB. In fact, those structures will determine the parameters of the SMC quite precisely. However, the cost of such accuracy is the need to solve a set of several coupled Boltzmann equations. For now, it is seen that there is a residual low level of ionization remaining well after decoupling. The observed acoustic structures in the CMB must wait for an exposition of inflation and the perturbations due to GR metric fluctuations and quantum fluctuations.

The Boltzmann equation used here is

$$y = (m_e c^2 / kT), \quad \Gamma \sim n_b(\sigma v), \quad H_{MD} \sim T^{3/2}, \quad X_e^o \sim e^{-B/2kT} y^{3/4}$$

$$dX_e/dy = -\frac{[X_e^2 - (X_e^0)^2]}{[(\Gamma/H)y^2]} \tag{5.3}$$

The difference in the $X_e = n_\gamma/n_b$ value from the equilibrium value X_e^o drives the evolution of X_e as a function of temperature. The effect of the Hubble expansion on the CMB decoupling through the Γ/H ratio is also accounted for now. The solution is numeric, using the "ode45" utility, as shown in Fig. 5.10.

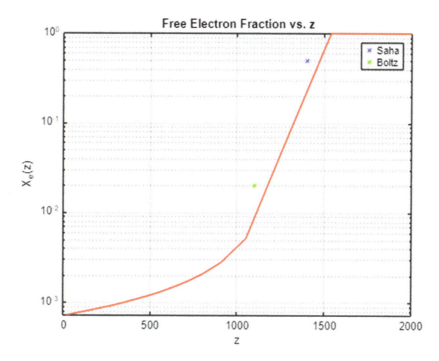

Figure 5.10: Free electron fraction near decoupling as a function of z. Dots are the Boltzmann (green) and Saha (blue) z for decoupling.

```
% look at CMB decoupling using the ~ complete Boltzmann eq
% constants
    h = 0.73 ;  % Hubble factor
    kb = (1.0 ./40) .*(1.0 ./300);  % Boltzmann k in eV/oK
    To = 2.73 ;   % CMB temp today (K)
    B = 13.6 ;  % H binding energy (eV)
    me = 511000.0 ; % electron mass in eV
    ngamo = 4.11 .*10 .^8 ; % photon # density 1/m^3, normalize CMP
-> Temp
omegb = 0.05; % Present Baryon Density/Critical
eta = omegb .* h .*h .*2.7 .*10 .^-8; % baryon to photon ratio
Tsaha = 0.26;   %(decouple in eV);
```

Xsaha = 0.0043

```
% Electron Ionization Fraction for Saha, 0.26 eV
% now look at dependence on final ionization on Omegab
```

Electron ionization at $z = 20$ is shown as a function of the baryon Ω in Fig. 5.11. The Saha estimate of the residual ionization is rather less than the Boltzmann estimate. For $z = 1000$, the Saha value is 1.16×10^{-4} while the Boltzmann estimate is ~ 0.005.

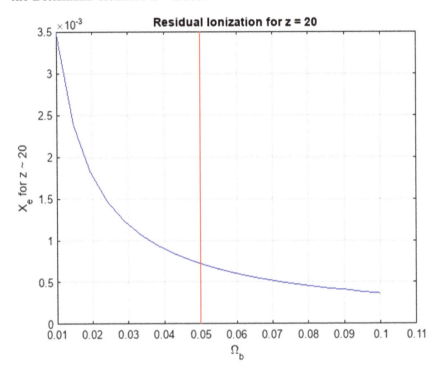

Figure 5.11:   Free electron ionization well after decoupling, $z = 20$, as a function of Ω_b.

5.4   Neutrino Mass

There are three types of charge neutral neutrinos in the SMPP, partners of the three types of charged leptons: electrons, muons, and tau leptons. Initially assumed to be massless, the neutrinos are now known to possess mass. Therefore, they should be part of the GR mass driven history of the Universe. However, the CP violation for neutrinos and the mixing among the three types is not yet well measured. Recalling that CP violation is necessary for baryogenesis, and that quark CP violation is insufficient, it is clear that improved measurements of the mixing and CP violations of neutrinos are of great importance. Indeed, there are several new experiments being constructed to address these specific issues. However, in this text, a simple method is used to estimate the possible range of impacts on cosmology as new neutrino data will arrive.

In this text, given the present uncertainties, the treatment of neutrinos is quite rudimentary. The number of relativistic spin degrees of freedom, g, is 2 for the massless photons and 1 for each of six neutrinos (three generations, particles and antiparticles). For bosons, the energy density is $(\pi^2/30)gT^4$, while fermions have an additional factor of 7/8. Similarly, there is an additional factor of 7/8 for fermion number density with respect to that for bosons. The effective temperature of photons and neutrinos also differs. Taken together, the energy density of relativistic neutrinos relative to photons has a factor of 0.68 which has been invoked previously:

$$T_\gamma/T_\nu = (11/4)^{1/3} = 1.40, \quad n_\nu = (3/11)n_\gamma$$
$$\rho_\nu/\rho_\gamma = [3(7/8)(T_\nu/T_\gamma)^4] = 0.68 \tag{5.4}$$

The number density for each of three generations of neutrinos at present is $(3/4)(4/11)$, or $n_\nu = (3/11)n_\gamma$, which is a substantial fraction of the photon number density, as was assumed for the nucleosynthesis treatment. Assuming there are only three distinct generations of massless neutrinos, the number density fraction of relativistic species would be multiplied by a factor of $1 + 3(3/11)$ or 1.82 and the energy density by a factor of 1.68. This is the origin of the factor 1.68 as an attempt to account for the neutrinos as a source term. In the text, T always refers solely to the photon temperature.

Since some neutrino generations have mass, they are presumably not relativistic at present. If the masses of all neutrinos were zero, then the "radiation" density would increase which would increase the scale factor at radiation/matter equality and decrease z at equality. Note that teq has been evaluated ignoring any contributions by neutrinos although the option of including three generations of massless neutrinos is available in the evaluation of the cosmological parameters by application of the factor 1.68. Although neutrinos have been ignored, their effects can be bracketed using the numerical solutions and changing Ω_γ to $\Omega_{\rm rcl} = 1.68\,\Omega_\gamma$ for the extreme case where there are three generations of massless neutrinos (Section 4.3). In that case, the value for $z_{\rm eq}$ goes from 6780 to 4042 and $T_{\rm eq}$ from 1.6 eV to 0.95 eV. Similar values can be found in texts which assume by default that the neutrinos are massless.

Neutrinos have a mass. Therefore, at late times, they will become non-relativistic. The experimental neutrino mass limit is presently 275 meV with different limits shown in Fig. 5.12. The limits on the lightest mass

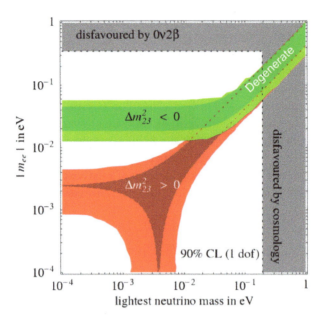

Figure 5.12:   Current limits on the lightest neutrino mass.

come from laboratory measurements of the end point spectrum of beta decays. The limits on the sum of the neutrino masses for three generations of neutrinos arise from studying the formation of LSS in cosmology as mentioned later in Section 8. The horizontal bands indicate estimated limits from present experiments, for example neutrino-less double beta decays, and expectations of future experiments:

$$\Omega_\nu \sim m_\nu(\text{eV})/50 < 0.0055 \qquad (5.5)$$

Neutrino oscillation experiments have shown that neutrinos have mass differences between $0.05\,\text{eV}$ and $0.009\,\text{eV}$. Therefore, primordial neutrinos are probably non-relativistic today, but the overall mass scale, not the differences, only exists as fairly weak limits at present. At t_{dec} with a temperature of $0.26\,\text{eV}$, they may well be a relativistic species. The neutrinos are NR at low temperatures and, UR at high temperatures. The "temperature" refers to that of the photons, however, which are bosons. There is a large relic of primordial neutrino abundance. There are direct experimental limits on the neutrino mass for double decay experiments, $<0.3\,\text{eV}$. There are limits arising from the measurements of flavor oscillations, which depend quadratically on the mass differences, leading to a twofold ambiguity. Cosmological limits also set a limit, in this case, $<0.2\,\text{eV}$. The "normal" ordering (red band) follows that of the quarks, while the "inverted" ordering (green band) is the second possible solution.

What about DM? It could have both weak interactions and mass. Sterile neutrinos, as yet only postulated, could explain ν masses and give a DM candidate with lepto-genesis of baryons but at mass scales which are probably unreachable by direct means. The amount of CP violation of the neutrinos needed to drive lepto-genesis is a crucial datum to measure. Future accelerator — based experiments are in preparation. A similar mixing matrix for the quarks has already been explored and measured quite accurately. For quarks, it appears to be unitary but with complex CP violating elements. In fact, CP violation is needed to explain the existence of matter so that the observation of CP violation for quarks is welcome. However, it seems to be of insufficient magnitude to explain the existing matter. The experimental program to map out CP violation for neutrinos is very important because sufficient CP violation is needed to explain the very existence of matter within the context of the SMPP.

The script below makes a cartoon, Fig. 5.13, of the evolution of matter, radiation, and neutrinos. The photon mass density scales as T^4. The neutrinos at higher T become relativistic. The abrupt transition shown in Fig. 5.13 is quite schematic. The user chooses the neutrino mass and thus the transition of the neutrino from non-relativistic to relativistic. Since neutrinos only interact weakly, their impact on the topics covered in this text is typically small and largely ignored because precision modeling of the neutrino effects is not yet possible. However, for precision cosmological models, neutrinos are of crucial importance. The SMPP measurements being readied will sharpen the SMC constraints and are a good example of the synergy between the two standard models.

```
% look at neutrinos in HBB cosmology, ignore vacuum energy density.
% radiation and matter domination only, HBB
mv = 200; % Enter Neutrino Mass in meV
% constants
% project back to equal densities - boundary of radiation/matter
Teq = Eo ./aleq   % T(eV) at photon/M Equality
```

```
Teq = 1.5445
```

```
% present matter, radiation and DE fractions of critical density
```

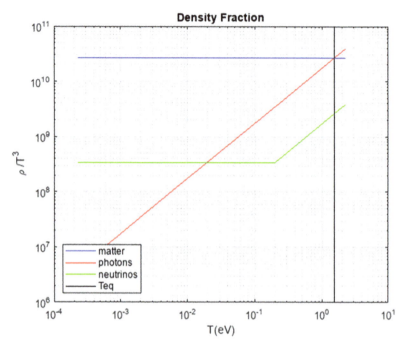

Figure 5.13: Schematic plot of the mass density of baryons, photons, and neutrinos as a function of temperature.

The photons rise linearly in the ρ/T^3 variable, while the matter is NR and constant at these temperatures. The neutrinos are NR at low temperatures but make a transition to UR at higher temperatures. A default neutrino mass of $0.2\,\mathrm{eV}$, the cosmological limit, is plotted.

5.5   Neutrino Masses and Oscillations

There are presently two solutions for neutrino masses. Since the neutrinos have mass, they can oscillate between their three "flavor" identities, ν_e, ν_μ, ν_τ, as do the quarks, u, d, s, However, the measurements of the mixing of identities in time depend on the square of the mass differences, leaving, for now, insufficient information to uniquely solve for the masses themselves. There are two possible quadratic solutions for the masses. The "normal" solution looks like the pattern observed for quarks, while the "inverted" solution does not. The overall scale, m_o, also needs to be well determined. Future experiments at new neutrino detectors will address the choice of the correct solution and better determine the elements of the neutrino mixing matrix, including the complex CP-violating terms which are important for baryogenesis scenarios.

In the first approximation, the neutrinos at the time of CMB decoupling are a relativistic, weakly interacting neutral fluid, which does not make a big impact on the CMB photon — "baryon" electromagnetically interacting fluid. The DM is assumed to only interact gravitationally, but weak interactions, like the postulated WIMP particles, are not precluded by the CMB data. The following script plots the three neutrino masses as a function of the basic mass, m_o for the normal and inverted possible solutions. The mass scales shown in Fig. 5.14 are $\sim$50 meV, chosen to be approximately lower than the current cosmological mass limits of $\sim$200 meV which are partially driven by the formation of large-scale structures later in the evolution of the Universe. In Fig. 5.15 the sum of the 3 masses is shown as a function of the mass scale.

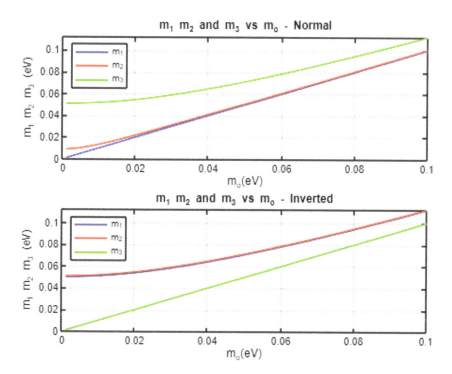

Figure 5.14: Plot of the three neutrino masses as a function of m_o for the normal solution (top) and the inverted solution (bottom).

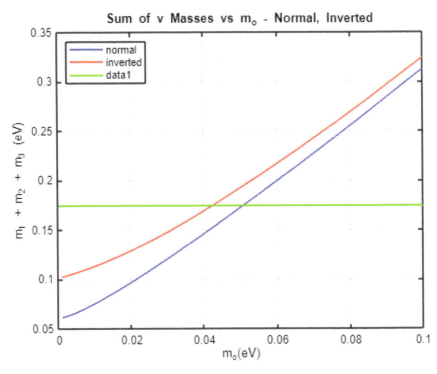

Figure 5.15: Plot of the sum of the three neutrino masses as a function of m_o. The green line is a recent limit at 95% CL on the sum of the three masses for the IH.

```
% neutrino masses, DBD, cosmo, solar osc, mixing of flavors 1,2, and
 3
m21sq = 8e-5;    % eV units
m32sq = 2.5e-3;
dmsq = m21sq;
dMsq = m32sq + m21sq ./2;
    mo = mo .*0.001; % lowest mass in eV
    m1N = mo; % normal hierarchy - quadratic ambiguity
    m2N = sqrt(mo .^2 + dmsq);
    m3N = sqrt(mo .^2 + dmsq ./2 + dMsq);
    sumN = m1N + m2N + m3N;
    % inverted hierarchy
    m3I = mo;
    m2I = sqrt(mo .^2 +dMsq + dmsq ./2);
    m1I = sqrt(mo .^2 + dMsq - dmsq ./2);
```

It is amusing that the CMB itself serves as a foreground target for ultra-high energy cosmic rays — protons by assumption. They interact inelasticly when pion production is above the threshold, and pions are produced with mb, $1\text{mb} = 10^{-26} \text{ m}^2 = 100 \text{ fm}^2$, cross-sections. The pion production threshold for a CMB target photon is at a proton energy of about 10^{13} GeV. With the current photon number density, the mean free path is then about 1 Mpc. This implies that cosmic rays incident outside the Milky Way will not reach the Earth. There is a cutoff, called the GZK cutoff for observing very energetic high-energy cosmic rays. A similar cutoff for high-energy neutrinos scattering off primordial neutrinos to produce Z bosons is not operational because the cross-sections are so weak that the cutoff mean free path is greater than or comparable to the horizon.

Chapter 6

From the Inflationary Big Bang

"Inflation challenges our intuitions about cause and effect,
because it suggests that the universe can create itself"

— Alan Guth,

"To be more childlike, you don't have to give up being an adult.
The fully integrated person is capable of being both an adult and
a child simultaneously. Recapture the childlike feelings of
wide-eyed excitement, spontaneous appreciation, cutting loose,
and being full of awe and wonder at this magnificent universe."

— Wayne Dyer

"They say the universe is expanding. That should help with the
traffic."

— Steven Wright

So far, the detailed structures in the CMB photons, which are inhomogeneous by only parts per 10^{-5}, have been ignored, and a perfectly spatially uniform black body spectrum has been assumed. This is a good first approximation but requires explication. There are issues of causal connections of the full, presently observable, CMB and of the smallness of curvature which have not been highlighted up until now. However, these are fundamental issues for the SMC to confront. This section explores the hypothesis of "inflation" as an explanation of these effects, without defining the source of the inflationary dynamics. The simplest initial approach is to postulate the early appearance of a Hubble parameter which is constant, such as a DE term would cause, $H = \sqrt{\Lambda/3}$. The size of H and the time over which it operates are specified by the two major problems which it is constructed to solve which concern the causally disconnected regions of the uniform CMB and the present flatness of the measured R–W metric.

6.1   Solving the CMB Causal Problem

In the HBB, the Universe has evolved over ~13.8 Gyr. The SMC defines the constituents of the Universe which drive its evolution from an HBB beginning at $t = 0$. The estimates for the time in the SMC when the photons decouple from the matter and begin to free stream are that this decoupling occurred, very approximately, at $t_{dec} \sim 400,000$ yr. The conformal time at decoupling was ~0.076. A plot is made below, Fig. 6.1, for the conformal time as a function of coordinate tine, and a second plot, Fig. 6.2, for conformal time as a function of conformal distance r at 10 times that value. That plot illustrates very different sections of the CMB could not have been in causal contact since the present conformal time is ~3.35. However, they have almost exactly the same BB spectrum and temperature, which constitutes a causality puzzle. Inflation is a scenario to solve the issue by placing the Universe in causal contact at early times. The user defines the range of coordinate time to be evaluated, and a full, fine grained numerical integration is performed with the SMC parameters.

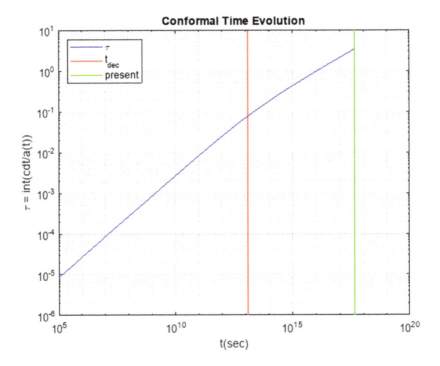

Figure 6.1:   Conformal time as a function of coordinate time.

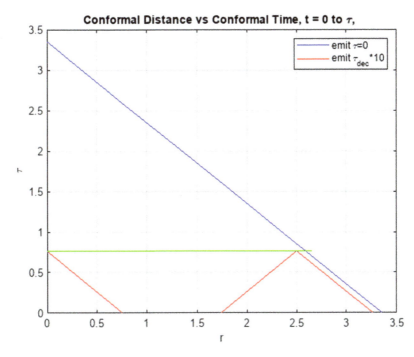

Figure 6.2: Plot of conformal time light cones as a function of comoving distance, r with τ scaled up by 10x.

```
% the horizon problem in Big Bang cosmology
% constants
% user chooses tmax and timin
tmax = 17.64 ; % Maximum Energy Exponent in sec (=17.64, present)
tmin = 0 ; % Minimum Energy Exponent in sec (E~MeV @ 1 sec,)
curv = 1.0 ./(Ha .^2);   % curvature in BB cosmo, |omega-1|/k = 1/(
Ha)^2
tion  % Time When Radiation Energy = H2 Ionization (yr)
```

```
tion = 308.1112
```

```
teq  %  Time When Radiation Density = Matter Density (yr)
```

```
teq = 1.8163e+04
```

```
tdec % Time When Radiation Energy = 0.26 eV = decoupling (yr)
```

```
tdec = 4.0378e+05
```

```
tau(idec) % Conformal Time When Radiation Energy = 0.26 eV =
decoupling
```

```
ans = 0.0758
```

```
tau(kk)  % Conformal Time at to
```

```
ans = 3.3460
```

```
tau(irm)  % Conformal Time at teq
```

```
ans = 0.0188
```

```
y(irm)  % Scale Factor at teq
```

```
ans = 1.4751e-04
```

```
% schemtic of causality for the CMB
% conformal time is the horizon ~ cto, in RW ~ c/Ho, if MD 3c/Ho,
explicit
% SMC calculation 3.35 DH = 14.8 Mpc
% SMC calculation for decoupling a t z ~ 1400, taudec = 0.076
```

The causality issue is not a minor issue. The comoving distance for the CMB photons is 0.076, while at present, the distance is 3.35. This is a causal mismatch by a factor of $\sim$44. There are $\sim$44 causally disconnected patches of the Universe visible today. Objects with comoving separations $> \tau$ today were never in causal contact. The causal problem can be solved if there is a limited inflationary period when a small, causally connected, smooth patch of space is inflated. The patch was causally connected before inflation, was driven outside the horizon by an exponentially increasing scale factor, $a(t) \sim e^{Ht}$, and forms the presently observable Universe, homogeneous and isotropic. A rough estimate of the inflation parameters can be made using, $N_e = 60$, the number of exponential factors during inflation, ending at t_e, with a constant H and $t_e = 4.7 \times 10^{-35}$s to solve the problem. The scale needed now to have a conformal time growth during inflation of $\sim$3.35 in order to just solve the causality problem is then $a(t_e) = cN_e/H$, $a(t_e) = 0.008$ m, or a scale $\sim$1 cm at the end of inflation. These estimates for t_e, H, and $a(t_e)$ set the approximate scales for a more realistic model of inflation.

6.2 Solving the Metric Flatness Problem

A second major issue in the SMC is called the flatness problem. If the Universe is not spatially flat, as was previously assumed in conformance with the present data, the radial term in the R–W metric is modified by a non-zero curvature parameter κ to be $d^2r/(1 - \kappa r^2)$. The κ parameter may be 1, 0, or 1. The current measured value of the flatness parameter is that the Universe is within a few percent of the critical energy density which divides positive and negative curvature so that κ is near zero. The evolution of the scale factor $a(t)$ is modified in the presence of non-zero curvature. This issue was briefly already mentioned in Section 3.6:

$$H^2 = \left(8\pi/3M_p^2\right)\rho - \kappa/a^2$$

$$\rho_c(t) = 3\left[H(t)M_p\right]^2/8\pi \qquad (6.1)$$

$$\Omega(t) = \rho(t)/\rho_c(t), \quad |\Omega(t) - 1| = \kappa/[H(t)a(t)]^2$$

The value for κ is defined in terms of the Hubble parameter, the energy density, the scale factor, and the Planck mass. In order to get a feeling for the effect of curvature, one can choose $H_o t_o = 1$ and a time variable x scaled to t_o, $x = t/t_o$, which yields an equation for the scale factor, $(da/dx)^2 = [\Omega_m/\alpha + \Omega_\gamma/\alpha^2 + \Omega_\Lambda \alpha^2 + (1 - \Omega_o)]$. In a Universe with only curvature, $a(t) \sim x$ which rises faster with time than either matter or radiation but not as fast as the exponential rise when there is substantial DE. The effects of curvature become less important at earlier times, while DE will dominate at late times.

The following MATLAB script solves the general problem numerically. The user inputs the DE and matter terms. Flatness is not necessarily assumed but depends on the inputs chosen by the user. The flatness parameter is defined by the behavior of $(Ha)^2$. For vacuum energy domination, H is a constant and the scale factor is exponentially increasing. Therefore, a vacuum energy density would drive that term rapidly to zero. In the HBB past, the curvature, $(\Omega - 1)/\kappa$, increased by a factor roughly 10^{15} from a time of one second to the present. Therefore, inflation solves the curvature problem by flattening the space enormously at early times, even though it would increase somewhat after inflation ceases to be effective in an RD or MD epoch.

So far, in the SMC, curvature was assumed to be 0. However, this is really an experimental question. The data limits the possibilities of curvature. For example, the script below shows the effect on $H_o t$ for a curvature as a function of z, as displayed in Fig. 6.3. The user inputs the DE and

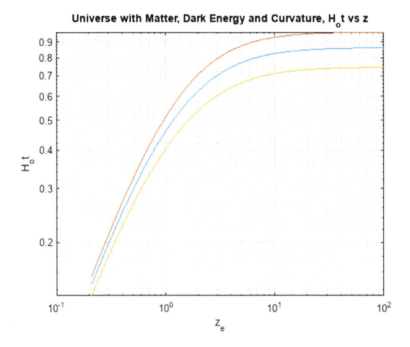

Figure 6.3: Evolution of $H_o t$ for reception at $z = 0$ from emission at z_e.

matter terms, while the negligible radiation term at late times is fixed at the SM value. The curvature term has the largest effect on the time development of the scale factor at early times. Present data put strict limits on the curvature factor of a few percent.

```
% find Hot for Universe with curvature
% implicit solution as a function of z
%  ignore radiation density, omk = 1 - omm - oml
omm = 0.2 ; % Enter Omega Matter
oml = 0 ; % Enter Omega DE
omk = 1.0 - omm - oml    % signed overdense/underdense
```

omk = 0.8000

```
% do the numerical evaluation
Hodt_dz = - 1.0 ./((1+z) .*sqrt(omm .*((1+z) .^3)+oml + omk .*((1+z)
  .^2)));
% lookback time is integral from z = 0 (to) to z = ze (te), emission
```

Inflation also solves the curvature problem, that is, the curvature is now zero within measurement errors. Intuitively, a vast inflation of the metric would smooth out any initial local curvature. Imagine a sphere stretched dimensionally by a factor 10^{26}. Any patch of the sphere would be locally flat. The present Hubble distance is $D_H = 1.37 \times 10^{26}$ m. A pure cosmological constant has a constant H, $H^2 = \Lambda/3$. Assume that the present time has a scale factor, $a_o = ct_o$, as was found to fortuitously be 0.95. For a constant H era, the conformal time is $\tau_\Lambda =\sim c/aH$. For a power-law behavior, $\tau = ct_o/[a_o(1-n)] = 1/(1-n)$, which is three for a purely MD epoch. In this case, the horizon is $3D_H$. Numerically, for the SMC, the value is ~ 3.35 times D_H.

In the presence of curvature, κ, and a cosmological term providing a constant H, $|\Omega(t)-1| = \kappa/D_H^2 = \kappa/[H(t)a(t)]^2$. The exponential growth of $a(t)$ rapidly drives the curvature of the space to be flat, $a(t) \sim a_o e^{Ht}$. The constant H is then required to make the present value of Ω to be 1, within the present experimental limit. Accounting approximately for the constant H phase followed by a growth in $a(t)$ in the HBB until the present, $\Omega_o - 1 = 0.01 = 1/[H(t)a(t)]^2 \sim (\tau_e e^{-Ht_e})^2 \left(\frac{t_o}{t_e}\right)$. Taking $a_o = D_H = 1.37 \times 10^{26}$m, the scale factor at the end of the constant H epoch is $a_e \sim 0.4$ m. The number of "e folds", is $N_e \sim 60$ (the log of 10^{26} is 62) and the growth of the scale factor from the end of the constant H epoch until the present is $a_e/a_o \sim \sqrt{t_e/t_o} \sim 10^{-27}$. Specifically, if $t_e \sim 10^{-36}$ s, $N_e = Ht_e \sim 65$, $a_e/a_o \sim \sqrt{t_e/t_o} \sim 2.2 \times 10^{-27}$. This sets the approximate scale of the growth of the scale factor during inflation which is necessary to solve the flatness problem.

To summarize, the issues for BB cosmology are the observed CMB uniformity and the resulting causality issue. In addition, the presently observed flatness of the metric is in contradiction to the curvature growth in both the RD and MD epochs. The numerical integrations already made with DE contributions at late and future times have pointed to a solution. The effect of DE at late times shows a rapid growth of $a(t)$ and a rapid decrease of curvature. Therefore, some form of an effective cosmological constant could solve the HBB problems while still retaining all its successes as long as it acted only at very early times. The constant H exercise is good for order of magnitude estimates and to convince oneself that inflation could solve the issues raised by the CMB uniformity and the lack of present curvature. However, simply switching on and off a massive H is only a first step. A dynamical field that naturally dies away is needed.

In addition, the inflation has cooled the Universe much like an expanding gas cools, and a mechanism to both end inflation and then reheat the Universe to retain the successes of the HBB is needed.

Conformal time with explicit curvature alone:

```
syms c r k rh tauk
k = 1;
tauk = (1/c)*int(1/sqrt(1-k*r*r),r,0,rh)   % conformal time with only
 curvature ~ exponential behavior
```

tauk =

$$\frac{a \sin (rh)}{c}$$

6.3   A Constant H Model

The HBB model has been a reasonable guide to much of the actual data. However, the details of the CMB cannot be accommodated without altering the model. The CMB is very uniform in all directions, even though remote sections of the CMB could never have been in causal contact with the HBB. A model called inflation was created to explain these facts in terms of an unknown field which caused the Universe at very early times to expand sufficiently to allow the presently observable Universe to be causally in contact at very early times. The expansion also flattened the metric, explaining why the present Universe appears to have zero curvature — it got stretched out. Subsequent curvature growth in RD and MD epochs is insufficient to create an experimentally measurable non-zero curvature.

The HBB model has been a success for results such as nucleosynthesis and, indeed, the SMC with source terms due to matter, radiation, and a cosmological term, albeit with an *ad hoc* dark matter and dark energy, both of which are presently without candidates in the SMPP. However, the detailed CMB properties cannot be explained. Not only the causal and flatness CMB issues, but also there are faint structures in the CMB photons on present scales about one degree that need to be understood. These structures are also evident in the clustering of galaxies as observed in astronomical galaxy surveys which provide a consistency check. The following sections address these issues. First, a constant H model that starts at $t = 0$ and ends abruptly is explored in more detail to set the needed scales.

The scale of inflation which is needed to solve the CMB issues has H energies of order 10^{14} GeV, temperatures $\sim 10^{27}$ °K, and a schematic of this model is shown below in Fig. 6.4 The energy scales are sufficiently large that gravitational waves need to be invoked and understood. Moving to a dynamic quantum field of inflation instead of a constant H as a model means that quantum fluctuations in the field are imprinted in the matter and photons of the CMB, imparting structures which remain after the CMB decouples from the matter. In this section, the constant H "model" is first used to set the scales and then a dynamic scalar field is invoked that slowly "rolls" with a large potential which decays away in order to end inflation. It ends with a cold Universe, so a very schematic reheating scenario is invoked in order to reignite the HBB and recover its successes.

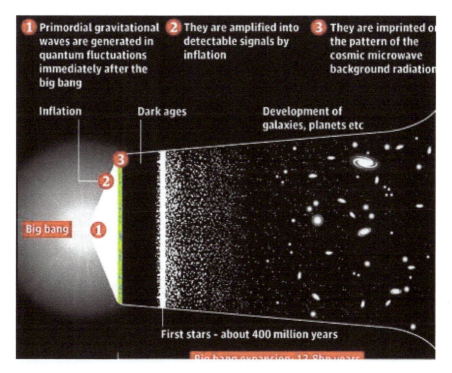

Figure 6.4: Inflationary period and HBB. Quantum fluctuations are amplified by inflation into structures visible today in the CMB and in matter.

Regions of the Universe at a time $\sim 10^{-34}$ s, spatially causally connected over a distance, $ct \sim 10^{-26}$ m, with radiation causing the growth of $a(t)$ would only now be ~ 1 m without inflation, $a_o/a(t) \sim T(t)/T_o \sim \sqrt{t_o/t} \sim$

10^{26}. An inflation factor of that order of magnitude is thus needed. A constant value of H is the simplest starting point. The user picks the initial value of H, $H = \sqrt{\Lambda/3}$, which, for now, is simply asserted to appear and then disappear. This constant H will drive an exponential growth of $a(t)$. The number of "e folds", as the jargon goes, needed to satisfy the requirements of inflation can then be considered in order to get an approximate idea of the magnitudes involved. For $N_e = 65$ "e folds", H is $\sim 10^{13}$ GeV. For fields, $\phi \sim M_p$, and an initial time scale $t \sim 10^{-38}$ s, as seen in the constant H scenario, the scales for a scaler field, $\phi, \phi t$, is $\sim h(10^{-5})$ and quantum fluctuations are expected. In what follows, the Planck constant is normally explicitly employed because it indicates the quantum nature of the effects. About the units, H has units of inverse time. Planck's constant is needed to convert to energies, $\hbar H$, which are more appropriate for discussions of inflation. The Boltzmann constant can then convert to temperatures. $\hbar = 6.58 \times 10^{-16}$ eVs, $k = 8.62 \times 10^{-5}$ eV/K°. In this text, sometimes units with $\hbar = c = k = G = 1$ are used in equations for simplicity, but the numerical results are always shown in MKS units.

The only parameter is H, which is chosen by the user. The end of inflation is calculated as such to drive the initial curvature low enough so that the present curvature is $<1\%$. After this inflationary period, assume radiation domination until t_{eq} at $\sim 5.7 \times 10^{11}$ s. Since $(Ha)^2$ goes as t in this period, $|\Omega_{eq} - 1| \sim 5.7 \times 10^{-4}/t_e$. From t_{eq} until the present at t_o, the scaling goes as $t^{2/3}$ leading to a present value of $|\Omega_o - 1| \sim 4.7 \times 10^{-37}/t_e$. Assuming that the present experimental deviation from flatness is only 1%, the scale for the time of the end of inflation is $t_e \sim 4.7 \times 10^{-35}$ s, in order to just solve the flatness problem. H and t_e define N_e and the scale at the end of inflation a_e. Plots, Fig. 6.5, are then made for a variety of constant H values. Although the number of "e folds" has a small range, the time of inflation varies over 3 orders of magnitude in this exercise.

```
% Look at constant H inflation
% pick a value for H and find te, N and ae
ho = 10; % Enter H in Units of 10^13 GeV
% Result for te in 10 .^-36 sec Units
H = ho .*1.5 .*10 .^37; % H in sec-1 units
tendinf = (x) abs(x - exp(-30*ho*x)*4.9*10 .^56);
% starting value
xend = fminsearch(tendinf,4);
te = xend .*10 .^-36 % sec
```

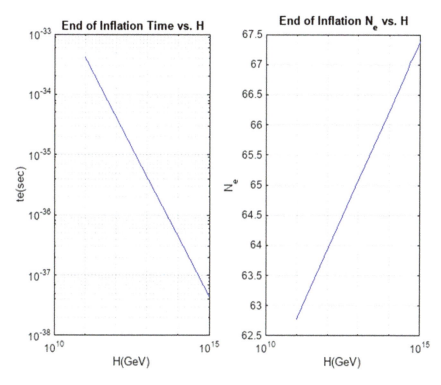

Figure 6.5:   End of the time of inflation (left) and number of total "e folds" (right) as a function of H.

```
te = 4.3789e-37
```

```
% End of Inflation (sec), Start at  t = 0 ; N e folds
Ne = H .*te
```

```
Ne = 65.6836
```

```
ao = 1.37 .*10 .^26 ; % present scale a in m assumed to be cto
to = 4.58 .*10 .^17 ; % to in sec
ae = ao .*sqrt(te ./to) % scale at end of inflation, m
```

```
ae = 0.1340
```

```
% End of Inflation, a(m)
```

6.4   Field, Potential, and Slow Roll

The constant H "model" solves the CMB causality and the flatness issues of the SMC with $\sim$60 "e folds", N_e. However, this is completely *ad hoc*; an enormous energy appears, persists for an infinitesimal time, solves the problems, and disappears. The next step would be to postulate some sort of dynamical scalar field, ϕ, that existed and then decays away. An inflationary field can be defined and the model compared to current experimental data. One can postulate a single potential and see if the value and time development can be derived and compared to cosmological data.

The scales of cosmological interest exit the horizon early during the inflationary period with N_e less than about 8 out of about 60. Quantum mechanics plus inflation naturally leads to small but observable density perturbations in the CMB which can be confronted with data. Inflation occurs at mass and timescales where quantum effects are small but presently observable. For fields of strength near to M_p and initial time of inflation of about 10^{-38} s where the product is $\sim 10^{-19}$ GeVs compared to which $\hbar$ is only $\sim 10^5$ times smaller which is about the size of the CMB fractional temperature variations.

The conformal time is $\tau \sim \int \mathrm{d}a/a^2 H \sim 1/aH$ if H is $\sim$a constant, or "rolls slowly". This is the comoving horizon during inflation. The comoving wave number would be $k = aH/c$. Assume the scalar field ϕ has a large initial potential energy, V, which would represent self-interactions, perhaps, as it does for the Higgs field. The potential needs to be approximately constant (slow roll) for long enough to solve the CMB causality and flatness problems. It then needs to speed up and fall to a minimum $\sim$0 at the "end" of inflation. A scalar field is needed because it must have the quantum numbers of the vacuum. As with the previous Higgs discussion, the field has dimensions of energy and the potential of energy density or energy to the fourth power. The field, ϕ, is assumed to be approximately uniform in space but decreasing with time. For now, there are no field fluctuations.

The present observable Universe, remaining after inflation, was a small homogeneous initial patch which was causally connected and inflated to become the homogeneous CMB. The Friedmann equations, Section 3.4, in the case of scalar fields with potentials are shown in the following. The energy density has kinetic and potential terms which are functions of the field ϕ. The relationship of ρ and p for scalar fields simplifies both the continuity equation and the acceleration equation. The acceleration of $a(t)$ is positive and inflation occurs if the potential term dominates since the pressure is negative. The Hubble parameter, energy density, time derivative (continuity equation), and scale factor acceleration for the field and potential are, in general,

$$H^2 = (8\pi/3M_p^2) \left[\left(\frac{d}{dt}\phi \right)^2 + V(\phi) \right]$$

$$\frac{d}{dt}\rho + 3H \left(\frac{d}{dt}\phi \right)^2 = 0$$

$$\left(\frac{d^2}{dt^2}a \right) \bigg/ a = - (8\pi/3M_p^2) \left[\left(\frac{d}{dt}\phi \right)^2 - V(\phi) \right]$$

$$\frac{d^2}{dt^2}\phi + 3H\frac{d}{dt}\phi + \frac{d}{d\phi}V = 0$$

(6.2)

If the potential dominates and varies slowly with time, $V(\phi) \gg \left(\frac{d}{dt}\phi \right)^2$, $\frac{d^2}{dt^2}\phi \sim 0$, H is quasi-constant and the acceleration is positive which ensures initial inflation. The field itself acts like a simple harmonic oscillator (SHO) with friction or damping due to H. The Hubble term supplies the damping since expansion makes smooth any oscillation. The restoring force is supplied by the derivative of the potential with respect to the field. The Hubble factor is proportional to the potential and the field decreases with time in a fashion depending on how the potential depends on the field:

$$H^2 \sim (8\pi/3M_p^2)V(\phi), \quad \frac{d}{dt}\phi \sim -(dV/d\phi)/3H \tag{6.3}$$

There are "slow roll" parameters that characterize the potential, V, through the first and second derivatives of the potential. As usual, a Taylor expansion of our ignorance is invoked as a first step beyond the constant H model, using the ϕ dependence of the potential V. These dimensionless parameters, ε and δ, characterize the first and second derivatives of the scalar potential and therefore approximately specify the shape of the potential. As is seen later, comparison with present cosmological data already imposes some constraints on the potential shape:

$$\varepsilon = (M_p^2/16\pi) \left(\left(\frac{d}{d\phi}V \right) \bigg/ V \right)^2, \quad \delta = (M_p^2/8\pi) \left[\left(\frac{d^2}{d\phi^2}V \right) \bigg/ V \right] \tag{6.4}$$

The scalar and tensor parameters are conventionally defined to be n_s, n_t with r the tensor to scalar ratio:

$$n_s = 1 - 6\varepsilon + 2\delta, \quad n_t = -2\varepsilon \tag{6.5}$$

The length of time for inflation to be active depends on the shape of the potential. The slow roll parameters are related to the time evolution of H and the field itself. It is only the acceleration of the scale factor which is proportional to H itself. Inflation ends when $\varepsilon \sim 1$ because at that point the acceleration of the scale factor has fallen to 0. Expressed in terms of the time dependence of H and the field the slow roll parameters are

$$\varepsilon = -\left(\frac{d}{dt}H\right) \Big/ H^2, \quad \delta = -\left(\frac{d^2}{dt^2}\phi\right) \Big/ H\frac{d}{dt}\phi + \varepsilon,$$

$$\left(\frac{d^2}{dt^2}a\right) \Big/ a = H^2(1 - \varepsilon) \tag{6.6}$$

In terms of these parameters, inflation starts with a small value of ε and ends when ε is one. Take $t = 0$ as the start of inflation and t_e as the end. There is an exponential growth of the scale factor during inflation. The number of "e-folds", N_e, is determined by ε:

$$N_e = \int_0^{t_e} H\,dt \sim \int_o^{t_e} 3H^2 d\phi \Big/ \left(\frac{d}{d\phi}V\right) = (8\pi/M_p^2)\int_o^{t_e} V/(dV/d\phi)d\phi$$

$$= (2\sqrt{\pi}/M_p)\int_o^{t_e} d\phi/\sqrt{\varepsilon} \tag{6.7}$$

The power in the fluctuations of the field is defined later. The power is treated as Fourier component, with a comoving parameter k. Since the time of horizon crossing of a Fourier mode, k, varies with mode since H is not strictly constant, the power is not scale invariant but varies with a "spectral index" n_s which introduces a factor into the power of $(k/aH)^{n_s-1}$. That factor is ~ 1 for the modes of interest, such as the CMB modes. In general, for future reference, the spectral index is

$$(n_s - 1) = (1/8\pi)(M_p/V)^2\left[-3\left(\frac{d}{d\phi}V\right)^2 + 2V\left(\frac{d^2}{d\phi^2}V\right)\right] = -6\varepsilon + 2\delta \tag{6.8}$$

In order to solve the issues with BB cosmology, it is expected that the number of "e- folds" should be ~ 60. Then the curvature would be driven very small, $1/(Ha)^2$, and the conformal time would inflate. The contribution to the growth of $a(t)$ depends on the shape of the potential. The number of "e-folds" depends on how fast the field is fractionally decreasing. The exponential increase in $a(t)$ drastically reduces the temperature since $T(t)a(t)$ is approximately a constant during inflation. After the field disappears, the Universe will need to re-heat to at least a sufficiently high temperature to create an HBB, with the light nuclei, whose relative abundance is predicted by HBB cosmology, and is a major success of the HBB model.

The physical mechanism for the existence of an approximately constant value of H which lasts for a limited time posits a scalar field which has a large initial potential energy and which depends on the potential shape such that it is sensibly constant (called "slow roll") for a sufficient period of time to solve the flatness and causal BB problems before it speeds up and falls into the minimum of the potential. The field then oscillates about its minimum decaying into less massive particles. The physics of the existence of the field itself and of the large initial potential, the shape of the potential, and the evolution of the field are unknown and are here simply assumed *ad hoc*. Many models are possible as long as they agree with the data. The tensor gravitational field is the only unique prediction of inflation, but it is not yet seen in the "BB modes" of the CMB. Current data results in values for the parameters of the potential V of $n_s = 0.965 \pm 0.004$ and $n_t = 0$ with an upper limit since the data are consistent with 0 within errors.

Inflation is the ultimate free lunch where the initial potential energy of a scalar field solves the HBB problems and gives, in addition, with its quantum fluctuations, a mechanism to predict and explain the richly structured Universe. This is the subject of the following two sections. This postulated field is not identified with a field known to the SMPP. However, the recently discovered Higgs field is a fundamental scalar field and is a possible candidate under the assumption of non-minimal coupling of it to gravity as is discussed later. This is a very interesting, compact, and economical possibility that "low energy" Higgs physics is an agent of Planck energy scale inflation physics. That would be the minimal assumption to attempt to connect the existing SMPP and the SMC. It is not, however, a "mainstream" model, and many theoretical models exist. In this text, a simple power-law potential is assumed, and the quadratic version is used as a numerical example. The script plots, in Fig. 6.6, $\phi(t)$ for several powers, b, of the potential from 1 to 3.95 (4 is unstable) assuming $V(\phi) = a^{4-b}\phi^b$ by symbolically solving the differential equation, Eq. 6.3, for the field:

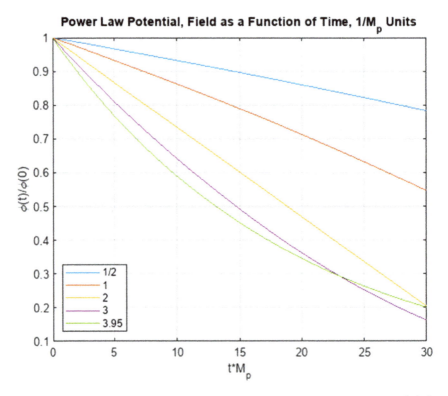

Figure 6.6: Time evolution of a scalar field with power-law potential, $V = a^{4-b}\phi^b$, $b = 1/2, 1, 2, 3, 3.95$.

```
% slow roll for the field as a function of time
% power law, single field, evaluate different exponents
    syms y(t) b c1 c2 Mp a
    c1 = b*Mp/(24*pi);
    c2 = c1*a^((4-b)/2);
    y(t) = dsolve(diff(y,t)== -c2*y(t)^(b/2-1),y(0)==1);
        % dphi/dt = -(b/24*pi*Mp)*a^(4-b)/2*phi^(b/2-1)
        % Symbolic Solution for phi(t) if V=a^(4-^b)phi^b
```

y(t) =

$$\frac{1}{\left(\left(\frac{b}{2}-2\right)\left(\frac{2}{b-4}+\frac{Mpa^2bt}{24a^{b/2}\pi}\right)\right)^{\frac{1}{\frac{b}{2}-2}}}$$

For the power-law potential, the time dependence of the field is found symbolically, printed, and then plotted above. A quadratic potential decays linearly, for example. For a power law field, ϕ^b, with the field chosen to yield N_e e-folds,

$$\varepsilon = b/(4N_e), \quad \delta = (b-1)/(2N_e), \quad r = 4b/N_e, \quad n_s - 1 = (2+b)/2N_e,$$

$$\phi(t_e)/M_p = b/4\sqrt{\pi} \tag{6.9}$$

For the special case of a quadratic field, $V(\phi) = m^2\phi^2/2$, as a worked example, the time derivative is a constant, and the field falls off linearly:

$$x = (\phi(0)/M_p), \quad \varepsilon = \delta = 1/(4\pi x^2), \quad 1 - n_s = 1/\pi x^2, N_e = 2\pi x^2 \tag{6.10}$$

In the previous constant H discussion, an H of 10^{32} GeV acting for 4.3×10^{-36} s yielded a $N_e = 64.5$ and a final scale of 0.42 m. The quadratic scalar field scenario is in very approximate agreement with these estimates even though it has a field which decreases linearly in time so that H also decreases during inflation and inflation stops. In the quadratic field model, $t_e = 10^{-36}$ s with a final field $\phi(t_e) = 0.28\,M_p$, the N_e value is 62.7, the final conformal time is 3.25, and the initial curvature is driven down by a factor 5×10^{-33}.

Solutions can be found for a general power law potential, $V = a^{4-b}\phi^b$. The field at the end of inflation, when $\varepsilon = 1$, is $\phi_e/M_p = b/4\sqrt{\pi}$. The slow roll parameters are $\varepsilon = (M_p/\phi)^2(b^2/16\pi)$, $\delta = (M_p/\phi)^2(b(b-1)/8\pi)$, with spectral index $n_s - 1 = (M_p/\phi)(-b(b+2)/8\pi)$. For example, for 60 e-folds and a quadratic potential, the field must start at $\sim3.1\,M_p$, $1 - n_s = 0.031$, $r = 0.13$.

For the CMB structures, in general, the potential for a single scalar field is normalized to be. $1.1 \times 10^{-6}V/\varepsilon = [2\pi(m/M_p)^2(\phi(o)/M_p)^4]M_p^4$. The quantity δ_H, the fractional quantum density fluctuation, for the magnitude of the acoustic structures, defines a CMB normalization $V/\varepsilon = (0.0054\,M_p)^4$. For the example of a quadratic potential, this requirement imposes the constraints that $(\phi(o))/M_p = 3.25, m/M_p = 1.1 \times 10^{-6}$.

6.5 Two Slow Roll Parameters

For the slow roll potentials of the power-law type, one can solve for the indices by constraining the prediction for the CMB power to be what is observed. This constraint on the slow roll models was already defined and is fully explained in later sections when the gravitational and quantum fluctuations are explored. For now, it is a normalization factor on the potential. The user chooses a specific exponent for a single power law, b, and the symbolic slow roll solutions are displayed. With a user-chosen power, the numeric values are then displayed. Plots are also then made for a range of exponents in Fig. 6.7, for the potential and the slow roll ε followed by plots in Fig. 6.8 for the slow roll variables, r and n_s. The required potential strength falls with the power-law exponent, whereas the ε parameter rises $\sim$ linearly with the exponent. In principle, then the strength of V and it's first two derivatives can be determined and characterized by the slow roll inflationary field model. For a slow roll model, the second derivative of the field is ~ 0, H is defined by the potential V, and N_e fixed by the derivative of the potential as in Eqs. 6.3 and 6.7, repeated here, $H^2 \sim (8\pi/3M_p^2)V(\phi)$, $N_e \sim \int H dt \sim - \int_{\phi_b}^{\phi_e} \sqrt{4\pi\varepsilon}(d\phi/M_p)$. The parameter a has a dimension of

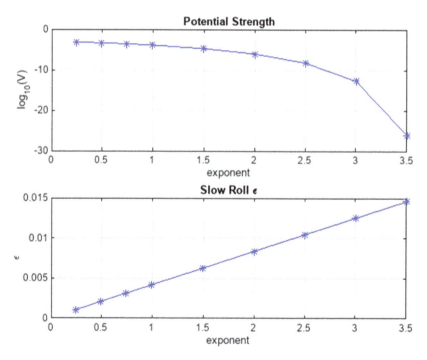

Figure 6.7: Plot of the strength of the potential V (top) and the ε parameter (bottom) as a function of the exponent.

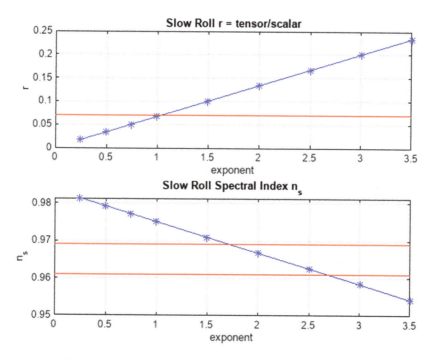

Figure 6.8: The r (top) and n_s (bottom) slow roll parameters as a function of exponent. Approximate experimental limits are in red. Taken at face value a simple power law is disfavored.

the energy such that V scales as the fourth power of energy:

```
% look at parameters of a power law potential
%  strength fixed by CMB power requirement for a given exponent
syms V eps1 eps del Mp x  alf a ns N1 b
V = a^(4-b).*x^b
```

$$V = a^{4-b}x^b$$

```
eps1 = (Mp*Mp)/(16*pi);
eps = eps1*((diff(V,x))/V)^2 % dervatives of V
```

```
eps =
```

$$\frac{\mathrm{Mp}^2 b^2 x^{2b-2}}{16 x^{2b}\pi}$$

```
del = (eps1*2)*((diff(V,x,2))/V)
```

```
del =
```

$$\frac{\mathrm{Mp}^2 bx^{b-2}(b-1)}{8x^b\pi}$$

```
ns = 1-6*eps + 2*del
```

```
ns=
```

$$\frac{\mathrm{Mp}^2 bx^{b-2}(b-1)}{4x^b\pi} - \frac{3\mathrm{Mp}^2 b^2 x^{2b-2}}{8x^{2b}\pi} + 1$$

```
bb = 2;
xi = sqrt((60 .* bb) ./(4.0 .*pi)) % initial field/Mp, 60 efolds
```

```
xi = 3.0902
```

```
P = V/eps; % power defines the solution
% get initial field using N, Mp = 1
% parameter a from CMB power - normalization
bbb = 0.0054 .^4 ;
% get initial field using N, Mp
```

```
epsss = 0.0083
```

```
etaaa = (bb .*(bb - 1)) ./(8 .*pi .*xi .*xi)
```

```
etaaa = 0.0083
```

```
nss = 1 - 6 .*epsss + 2 .*etaaa
```

```
nss = 0.9667
```

```
% ratio of tensor to scalar power
r = 16 .*epsss
```

```
r = 0.1333
```

There are both GR metrical fluctuations, δ_h, and inflationary quantum fluctuations of the field, δ_H, which are imprinted on the CMB and the structure of matter. The fluctuations are related by the slow roll parameter, $\delta_h = \delta_H\sqrt{9\varepsilon}$. The tensor power is then related to the scaler power as $P_t = P_s(9\varepsilon)$, $1/\varepsilon = 4\pi(\phi/M_p)^2$, with a tensor-to-scaler ratio of 0.12 in the case of a quadratic potential. The fractional quantum fluctuation of the scaler field δ_H is dimensionless and the power carried by that fluctuation P_H is defined. Note that these H subscripted variables are not the Fourier components which are not subscripted and which are used in later Chapters:

$$\delta_H = [(\hbar H^2)/(2\pi(\mathrm{d}\phi/\mathrm{d}t))] = \sqrt{128\pi/3}[V^{3/2}/M_p^3(\mathrm{d}V/\mathrm{d}\phi)]$$

$$P_H = \delta_H^2 = 8/3\left(V/M_p^4\right)/\varepsilon = (H/M_p)^2/\pi\varepsilon \tag{6.11}$$

In the specific example of a quadratic potential, $V = (m\phi)^2/2$, $\delta_H = \sqrt{16\pi/3[(m/M_p)(\phi(0)/M_p)^2]}$, $P_H = (2/3)(m/M_p)^2 N_e$, and the time derivative of the field is constant.

Limits on the slow roll parameters come largely from the analysis of the detailed structures in the CMB. How these arise is dealt with in later sections after a more detailed discussion of quantum fluctuations of the fields and gravity waves generated during inflation. For now, it suffices to show how the single power-law potential generates predictions for the slow roll parameters that are compatible with the experimental data. The plot on the left comes from data from 2015–2018 largely from the PLANCK collaboration. It was used to set limits on the slow roll inflation parameters.

For Fig. 6.9, the plot on the right is the 2022 PDG compilation, which shows the experimental progress in limiting the slow roll parameters. A non-zero value for the n_s parameter was well established and the tensor-to-scalar ratio was limited to be < 0.15. The current experimental limits are shown in Fig. 6.8 as red horizontal lines which favor exponents ~ 2 for the power-law potential in the simplest single field power-law potential model. However, the predictions for the tensor piece favor a somewhat softer exponent, favoring exponents, <1, although a definitive measurement has not yet been made, and only an upper limit exists. Indeed, this tension implies that a simple single field, power-law model is perhaps very much over-simplified. A power law is the simplest *ad hoc* assumption that can be made. Many other models of the potential exist. Time will tell.

In addition, new data from telescopic surveys of galaxies, beginning with the Sloan Digital Sky Survey (SDDS) and followed by the Dark Energy Survey Instrument (DESI), would provide a consistency check and, in addition, would improve on the parameter limits. There is a cross-check because the baryonic matter was tightly coupled to the photons until decoupling and therefore received the same acoustic oscillations as the photons. Therefore,

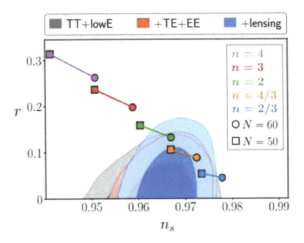

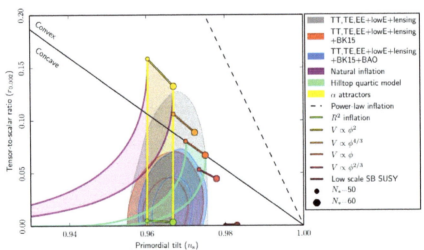

Figure 6.9: Experimental limits on the two slow roll parameters for a range of e-folds, N, and power-law exponents, n. Exponents $n = 4$ and 3 are disfavored (top), data compilation 2018 (bottom) 2022.

a cross-check is possiblly comparing the CMB acoustic peaks and the corresponding structures in the matter distributions.

Soon, the new B mode experiments will help better define r and place better restrictions on (n_s, r) and experimenters will begin to specify the shape of the inflationary potential. More accurate data will also limit higher-order derivatives of the potential. For example, the third derivative of the potential contributes to the evolution of the spectral index with k. $dn_s/dln(k) = 16\varepsilon\delta - 24\varepsilon^2 - 12\xi, \xi = M_p^4 \left[(dV/d\phi) \left(\frac{d^3}{d\phi^3} V \right) \right] / V^2$. For the quadratic field model $dn_s/dln(k) = -2/N_e^2 \sim -00055$.

6.6 Inflation, GR, and Quantum Mechanics

The regions of mass and time covered by inflation are much more extreme than those covered by the HBB model. For example, the fields needed in the slow roll model have masses comparable to the Planck mass. It seems clear that GR will be very important and that, in fact, gravity waves will be generated in the inflationary period. Quantum fields are needed, as seen in the regimes needed for the fields used to describe inflation and these fields will have intrinsic quantum fluctuations. The plot generated in Fig 6.10

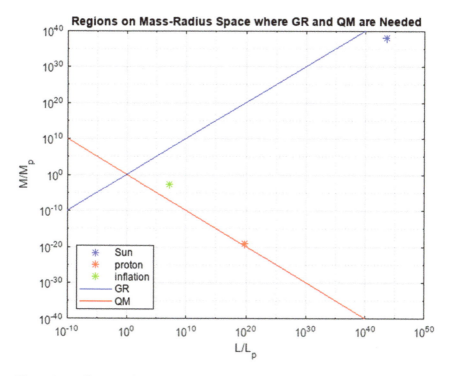

Figure 6.10: Cartoon of the regions in length and mass space where GR and quantum mechanics are needed.

is a cartoon simply to highlight the need when exploring inflation for GR gravitational radiation, and the need to have quantum fields, looking perhaps like a Higgs Lagrangian with kinetic terms and a power-law potential already invoked to drive slow roll inflation.

To set a scale for Fig. 6.10, the basic quantum parameter is $\hbar c$ which is 0.194 GeVfm, or $\hbar = 0.6 \times 10^{-24}$ GeVs. A quantum fluctuation with a scale of energy of 1 eV will occur inflated by a factor $\sim 10^{26}$. Quantum fluctuations will be stretched to macroscopic sizes by inflation and will be visible as a structure in the CMB. It will now be the next step to track the quantum and metric fluctuations from inflation to the CMB through the HBB. It seems a combination of elements of QM and GR will be needed to fully explore inflation, even in the absence of a full quantum theory of gravity. For a scaler field in the environment of a Hubble factor H, the field fluctuations are $\delta\phi \sim \hbar H/2\pi$.

```
% look at regimes where GR or QM are needed
% constants
    Mp = 1.2e19 ;   % Planck mass, GeV
    Rpl = 1.6e-35 ; % Planck length in m
    Ms = 1.1e57;    % solar mass (kg)
    Rs = 6.9e8 ;    % solar radius in m
    Mpr = 0.94 ;    % proton mass, GeV
    Rpr = 1e-15 ;   % proton radius, m
% quadratic inflation model
% V .^1/4 and L = ct_i
Vinf = 1.8e-3;   % V^1/4/Mp
Linf = 3e-28;    % L = ctinf
```

6.7   Slow Roll Solves HBB CMB Issues

The slow roll model improves on the constant H idea in that the initial field is a dynamical object which decreases with time during inflation. The initial field is, however, still arbitrary in magnitude and shape. For a slow roll potential, a simple power law of V can be tried as a worked example. A requirement is that the number of e-folds is ~ 60 in order to confront the flatness and causality issues in the CMB. The user picks the initial value of the field and the mass of the field. The potential V is fixed to scale as the square of the field, $V(\phi) = m^2\phi^2/2$, although other powers could be attempted. The metric is tracked at early times, where the curvature is assumed to be initially large. The field for the quadratic potential falls off linearly with t as does H, both falling to near zero at a time of 10^{-36} s. The curvature is greatly reduced, by $\sim$60 orders of magnitude, during inflation and then rises after the end of inflation occurring at $\varepsilon = 1$. However, at present, it is still very small as is consistent with observation. The conformal time rises during inflation and is within the conformal light cones of the HBB just after the end of inflation. Specific results for a quadratic potential as a worked example are

$$\phi\left(t_e\right) = M_p/\sqrt{4\pi}, \quad \varepsilon_e = 1, \quad \phi(o)/M_p = \sqrt{N_e/2\pi}, \quad \frac{\mathrm{d}}{\mathrm{d}t}\phi = mM_p/(12\pi\hbar),$$

$$\hbar H(t) = \sqrt{4\pi/3}(m/M_p)\phi(t) \tag{6.12}$$

In the following example, the initial field scaled to M_p is 3.1 and the potential mass is 1. The resulting final field and H as a function of t are shown in Fig. 6.11 while $a(t)$ and $N(t)$ appear in Fig. 6.12, where slow roll has ended, is $\phi \sim 0.3\, M_p$. Inflation ends at $\sim 1 \times 10^{-36}$ s.

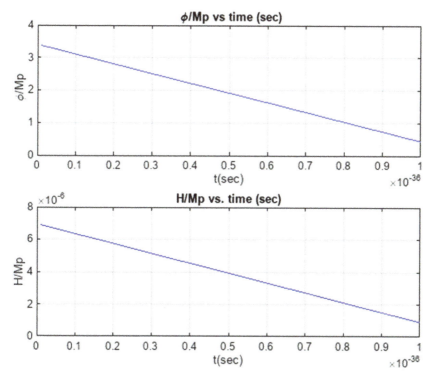

Figure 6.11: Linear time dependence of the field (top) and H (bottom) during slow roll inflation.

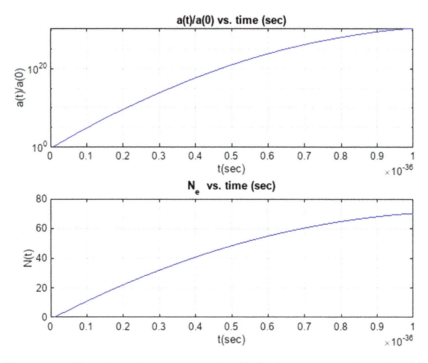

Figure 6.12: Time dependence of the scale $a(t)$ (top) and the number of e-folds (bottom).

```
% look at slow roll inflation as solution to flatness and horizon
problems
% constants
% inflation ends when epsilon ~ 1 or phie ~ Mp/sqrt(4pi)
phie_Mp = 1.0 ./sqrt(4.0 .*pi)
```

phie_Mp = 0.2821

```
% inflation starts with phii field - large enough to make N
expansions
NN = 60;
phiiN_Mp = sqrt(NN ./(2.0 .*pi))
```

phiiN_Mp = 3.0902

```
% Need Large Enough Initial Field for N ~ 60 e Folds
phii_Mp = 3.4 ; % Initial Field phi/Mp >3.1, N > 60
```

```
% simple scaler field - V = m^2phi^2/2
% H ~ (m/Mp)phi, a ~ exp(int (phi dt)), adot = Ha ~ a*(m/Mp)*phi
% slow roll at start of expansion defines the power spectral index
% Field mass defines time decay of Field = m/Mp~10^-6
% slope of roll = m/sqrt(12*pi) - defined by m => N
% field slope (1/t) = ddd and decay time (sec) = tff
m_Mp = 1 ; % scaler field mass/Mp x10^-6, defines final roll time
% units 1/sec, slope in t of V
```

dd = 2.9612e+36

te = (phii_Mp - phie_Mp) ./dd

te = 1.0529e-36

The full history of both curvature and conformal time, with inflation followed by HBB, appear in Fig. 6.13. Note that the curvature is assumed to start at some ill-defined large initial value. Note also that, strictly speaking, conformal time is negative during inflation, $d\tau = cdt/a = (c/H)da/a^2$, so that for constant H, $\tau = \int cda/a^2 = -c/(aH)$. The flatness parameter is defined by the behavior of $(Ha)^2$. For vacuum energy domination, H is a constant and the scale factor is exponentially increasing. Therefore, a vacuum energy density would drive the curvature rapidly to $\sim$ zero. This behavior was already seen when extrapolating to future times when the vacuum energy more fully dominates. In the past, the curvature increased by a factor roughly 10^{50} from a time of one second to the present. The expected linear dependence of the flatness parameter on t, $t^{2(1-n)}$, is observed for the RD epoch with a t dependence, followed by a softer $t^{2/3}$ behavior for MD times.

The problem is that if the curvature is now near zero, then in the past, it must have been incredibly fine-tuned to be almost exactly zero. The deviation from flatness is controlled by the square of the value of Ha. The curvature is not stable but grows with time. Therefore, a flat space is unstable as time increases and fine-tuning seems difficult to avoid as a necessary initial condition in the Universe. This is effectively the required initial condition for BB cosmology and inflation is needed as the initial condition. The value of conformal time during the very limited period in time of inflationary acceleration can also be comparable to the value for the entire later evolution of the Universe.

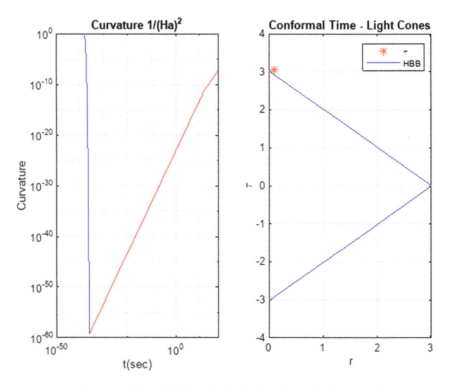

Figure 6.13: Curvature during (blue) and after (red) inflation (left), and conformal time at present (right).

6.8   Horizons in Slow Roll Inflation

The slow roll model gives more structure and dynamics than the constant H model. Although the initial field/potential is *ad hoc* and chosen by the user, it will decrease in a slow roll and inflation will end in an explicable fashion. Modes will be driven out of the horizon which will re-enter at later times either in an RD epoch after re-heating or in a later MD epoch. The timescale for generating the needed ~60 e-folds is ~10^{-36} s. The scale $a(t)$ grows by a factor ~10^{28} during inflation, followed by an assumed immediate re-heating into an RD epoch, followed by the MD epoch to the present. The "reheating" is discussed presently. The quadratic potential model is assumed as a concrete and simple example. The user chooses the initial value of ϕ. Conformal time is $\tau = c/(aH) = d_H$. The slow roll parameters for the quadratic potential were quoted previously, $\varepsilon = 1/[4\pi(\phi(o)/M_p)^2]$, with index, $n_s = 1 - 4\varepsilon$. The CMB temperature fluctuations are used to normalize the H fluctuations, $\delta_H = 2 \times 10^{-5}$. Details are supplied, as always, in the "comment" lines of the script:

$$H \sim (m/M_p)\phi, \quad \frac{\mathrm{d}}{\mathrm{d}t}a \sim Ha, \quad a \sim e^{(m/M_p)\int \phi \mathrm{d}t},$$

$$D_H = c/H, \quad d_H = c/aH \tag{6.13}$$

A word about notation is in order. In general, upper case will be used for physical quantities, and lower case for comoving. The two wave vectors are $k = aH/c$, $K = \alpha H/c$, where k is dimensionless and K has dimensions of inverse length. A dimensionless wavelength is $\lambda = 2\pi/k$. The one with dimensions of length is $\Lambda = 2\pi/K$. For example, $K_{eq} = 0.015$ Mpc^{-1}, $K_{dec} = 0.0045$ Mpc^{-1}, $K_o = 2.25 \times 10^{-4}$ Mpc^{-1}. Horizon crossing occurs at $k\tau = 1$, $\tau = c/aH$. Horizons are very important in the inflation paradigm, and various Fourier modes will pass out of the horizon and can re-enter at a later time. For this reason, Fourier analysis is used because the conformal quantities are constant during inflation, while dimension-full quantities are enormously stretched. The scale $k\tau$ is the Fourier comoving wave number times the comoving horizon. If it is <1, the wavelength is so big that there is no causal physics operating. In some sense, inflation goes faster than light. The constant wavelength can later, after inflation, re-cross the horizon and re-enter into causal interactions. A scale is driven outside the horizon and re-enters only much later, in the RD or MD epochs. The green line in Fig. 6.14 shows a mode driven out of the horizon early and re-entering near the RD/MD transition. Large scales, small K, large Λ, exit the horizon early and re-enter late in the MD epoch, while smaller scales, large K, small Λ, exit late and re-enter early, crossing in the RD. Due to a spread in horizon crossing times, the spectral index, n_s, was introduced. A constant H has a spectral index of 1.

The CMB temperature fluctuations are at a fractional scale $\sim 10^{-5}$, which sets a scale for the scalar field mass in the power-law model. A detailed structure of those fluctuations requires a full Boltzmann equation analysis and is deferred until later. For now, the quadratic field worked example is used to examine the horizon for different comoving Fourier modes. For both inflation and HBB epochs the scale $a(t)$ and the conformal time $\tau(t)$ are shown as a function of t in Fig. 6.14.

```
% slow roll inflation, coordinate distances, Hubble sphere, horizon
physical distances
% constants
    ao = 1.374 .*10 .^26;    % m - approx present scale factor, BB t
= to
```

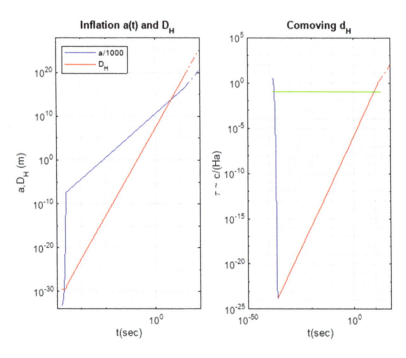

Figure 6.14:   Scale factor and Hubble distance vs t (left) and conformal time (right).

```
c = 3.0 .*10 .^8;        % m/sec
to = 4.58 .*10 .^17;     % sec - approx present time
teq = 7.6 .*10 .^11;     % time of equal rad - matter
Mp = 1.2 .*10 .^19;      % GeV - Planck mass
hbar  = 0.66e-24;        % GeV*sec
tp = hbar ./Mp;          % Planck time in sec
Lp = tp .*c;             % Planck scale in m
% use example of Quadratic Potential, Horizon
% inflation ends when epsilon ~ 1 or phif ~ Mp/sqrt(4pi)
phif_Mp = 1.0 ./sqrt(4.0 .*pi);
% inflation starts with phii field - large enough to make N
expansions
phiiN_Mp = 3.1;   % consistency - slow roll, N ~ 60
% simple potential with scaler field - V = m^2phi^2/2
% H ~ (m/Mp)phi, a ~ exp(int (phi dt)), adot = Ha ~ a*(m/Mp)*phi
% slow roll parameter at start of expansion defines the power
spectral index
epsilon = 1.0 ./(4.0 .*pi .*phiiN_Mp .*phiiN_Mp);
ns = 1.0 - 4.0 .*epsilon % epsilon = eta in slow roll
```

```
ns = 0.9669
```

```
% Fluctuation Power Spectrum Depends on phi/Mp Initial Value
% Fluctuation Spectral Index = ns, COBE data ~ 0.96
% COBE T fluctuations define scaler mass m/Mp ~ 10 .^-6
% Scalar Mass for CMB dT/T ~ m/Mp~10^-6
m_Mp = 10 .^-6;  % consistency with slow roll
% now scale with radiation dominated power law to
% approximate time for rad = matter densities
```

As displayed in Fig. 6.14, inflation greatly expands the scale $a(t)$. After exponential inflation, the RD (solid) and MD (dashed) epochs have a power law increase in $a(t)$ and $D_H = c/H(t)$. The conformal distance is driven very low by inflation, c/aH), outside the horizon, but rises as a power law in the RD and MD epochs and may re-enters the horizon during those times, as indicated by the horizontal green line. Larger Λ exits the horizon earlier and re-enters the horizon later. Conformal quantities are easier to visualize in regards to horizons as expected.

6.9 Inflation to HBB: Reheating

The slow roll power-law potential can generate $\sim$60 "e-folds", N_e, and thus perform the necessary work of inflation, to fix the issues of the HBB with causality and flatness. Although there is no consensus on the start of inflation, as the field rolls down, it ends at a time when $\varepsilon = 1$ and the acceleration of the scale factor is 0. Thus, an appropriately chosen potential will drive inflation and will naturally end the inflationary period. After that, the Universe, having been spectacularly expanded, is very cold and flat. In order to recover the successful HBB, a mechanism to "reheat" is needed. This process occurs at energies far in excess of those explained by the SMPP, by a factor $\sim 10^{16}$. This factor is unlikely to be dramatically reduced in the near future by the creation of new accelerators. Thus, the reheating scenario is not defined by SMPP physics, and speculative models can be constructed and compared to such data as exist.

After the explosive growth of the scale factor during inflation, additional growth is hypothesized in an MD epoch when the residual inflation potential converts to mass which ultimately decays into the lightest stable elements of the SMPP, electrons, neutrinos, photons, protons, and neutrons. At this point, the HBB can be considered to have been established.

One simple idea is that after the end of inflation, the field ϕ gets stuck in a false, non-zero, vacuum, or vev, begins to oscillate about the minimum, and creates particles of some sort. These are assumed to be unstable and ultimately decay to stable SMPP particles. The field disappears after converting its energy into SMPP particles with different masses. The sequence is thought to be inflation with exponential growth of the scale, an MD cold epoch re-heating and igniting an HBB, which then proceeds. At some point, baryons are created so that matter will exist, as well as DM. How and when this occurs is not known, a schematic view of the full history of the Universe appears in Fig. 6.15.

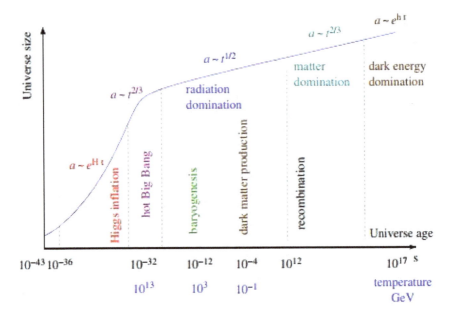

Figure 6.15: Schematic of $a(t)$ as a function of t and T. After exponential growth, an MD reheating epoch ignites the HBB.

A cartoon of the energy of the Universe as a function of the number of e-folds follows in Fig. 6.16. Since the physics which operates at these energies is unknown, it is of necessity very schematic.

```
% Movie of N (e folds) during BB
% from GUT, LHC, n, He, RD/MD, CMB, stars, today
% BB Movie of the Expansion History
```

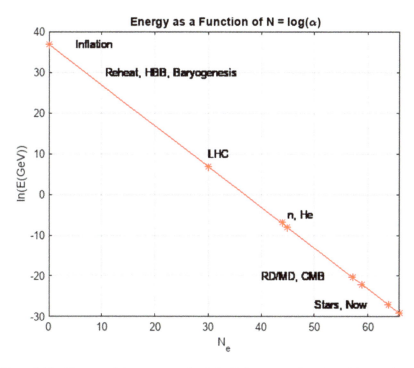

Figure 6.16: Cartoon of the energy scale of the Universe as a function of the number of "e-folds" since inflation.

```
% From GUT to WIMP to He Formation, to RD/MD Equality, to CMB. to
Stars, to Present
% use N as the "natural time marker, vs Mass scales which are
relevant
% track energy/temperature
```

There are ~30 decades in energy from the end of inflation until the highest laboratory energies reached today. Nevertheless, it is important to connect inflation smoothly to the HBB. The end of inflation has already been estimated for a power-law potential. In the absence of guidance, an effectively instantaneous conversion of the energy in the scaler field into matter and radiation can be asserted to occur and the consequences checked against all the relevant cosmological data. A radiation-dominated HBB is recovered and the HBB cosmology goes forward. Our knowledge has a serious gap between the SMPP and the scenario for inflation. Nevertheless, some crude models may be of some use and are be briefly explored.

6.10   Reheating and Reigniting the HBB

Rather than the previous simple assertion that inflation ends abruptly and the Universe reheats immediately in order to recover the HBB, a slightly more detailed description can be attempted, even though the physics at reheat energies is unknown.

Inflation ends, and the field $\phi, V(m, \phi)$, is still in a vacuum state with no particles. Then the vacuum decays into particles as it oscillates about some undefined potential minimum rather like the case of the Higgs boson VEV. These particles then in turn decay into SMPP particles. After the end of inflation, the residual field, in the vacuum state, can be thought of as a condensate of massive scalar particles with zero momentum. In some sense, the "latent heat" of the field causes an energy release of order $V(t_e)$ which approximately restores the temperature which was driven down by a factor e^{-N_e}. This non-adiabatic process leads to a large entropy increase, an entropy, for photons, which has already been invoked in the HBB nucleosynthesis discussion.

While the decay rate Γ is greater than H, there will be thermalization. The post-inflation temperature is small since T scales as $\sim 1/a$, and during inflation, the scale $a(t)$ grew enormously cooling the Universe. This low temperature drives the density of relic primordial states, for example, monopoles, to be very low thus inflating away any relic particles created earlier to levels consistent with present limits. This prediction is a success of the inflationary paradigm if monopoles or other relics were expected in a given theory, say a GUT model.

The reheat temperature must be high enough, early enough, to ignite the HBB and retain the HBB successes so that T should be at least ~ 1 MeV if not much larger. At the end of inflation, ε is equal to one and the energy density is still dominated by the potential, $V(\phi)$, which in the quadratic model with $m/M_p \sim 10^{-6}$ is $\rho_e \sim (6.4 \times 10^{15} \, \text{GeV})^4$. Note that the current critical energy density is $\rho_c \sim (2.6 \times 10^{-12} \, \text{GeV})^4$ by comparison. The decays must be rapid if high-temperature scales are to be established. How these decays occur is unknown in any detail, and the degrees of freedom of the fields to which the scalar field couples are speculative.

The problem with only photons is that there is matter in the Universe. Baryogenesis is necessary and can be accommodated at the GUT scale, for example, which is a desirable feature if not one which is yet experimentally established. In that case, the pre-heating time should be small. There are three necessary conditions for baryogenesis: a period out of thermal equilibrium, baryon-violating interactions, and CP-violating interactions. The present SMPP cannot accommodate baryogenesis and physics beyond the SMPP, such as exists in GUTs or higher symmetry groups than the SMPP, is needed. However, CP violation has been observed, and a period out of equilibrium is expected. An issue is the present-day non-observation of baryon number violation, for example, the failure to observe proton decay. A plausible baryogenesis model does perhaps not yet exist. In addition, in some fashion, DM must also be created, even leaving aside DE.

The scalar field, after the end of the rundown of the potential, generically acts like a damped oscillator moving harmonically around an assumed VEV while decaying:

$$\frac{\mathrm{d}^2}{\mathrm{d}t^2}\phi + (3H + \Gamma)\frac{\mathrm{d}}{\mathrm{d}t}\phi + m^2\phi = 0 \qquad (6.14)$$

The $3H$ factor is simply damping due to expansion as seen previously and the decay rate term adds to the damping. The remaining potential decays and the Universe is pre-heated and thermalized. In general, thermal equilibrium exists if the reactions are not smoothed by the expansion, as invoked already for WIMP searches, nuclei, and the CMB. Dimensional argument leads to a simple decay width $\Gamma(\phi \to x + x) \sim \alpha_m m$, which is proportional to the scalar decay particle mass, m, and some coupling constant, α_m, if the decays of the field are first-order two-body decays. The decay daughters, x, are assumed to be SMPP objects or at least objects with masses much less than the scalar mass m. Clearly, there most probably will be cascade decays down to SMPP particles. The contribution of produced particles to the total energy density is large when $3H$ becomes comparable to Γ. Applying the Stefan–Boltzmann law, with a number of degrees of freedom, $g^* \sim 100$ appropriate to the SMPP at high temperature, leads to a thermalized reheating temperature T_R and energy density:

$$\rho_{\text{re-heat}} \sim (\Gamma M_p)^2 / 24\pi = g_* \pi^2 T_R^4 / 30, \quad \Gamma \sim \alpha_o m \qquad (6.15)$$

The grand unified models (GUTs) imply a unified dimensionless coupling constant of about $1/24$ appropriate when the strength of the three SM forces come together at about 10^{16} GeV or about $10^{-3} Mp$. Solving for the reheating temperature, which is about a factor 30 larger than the scalar mass m, $T_R \sim 0.14\sqrt{\Gamma M_p} \sim 3 \times 10^{-5} M_p$. Extrapolating the HBB backward in time and scaling for radiation domination to 10^{-36} s, T rises to $\sim 10^{15}$ GeV compared to $T_R \sim 3.4 \times 10^{14}$ GeV. The extrapolated time for the GUT scale is 10^{-38} s. For the decays to m particles, H/Mp is $\sim 10^{-6}$ at the end of inflation and the ratio of $3H/\Gamma$ is ~ 71, which means that decays will not dominate over H until a time of 7.1×10^{-35} s, which is not consistent with rapid decay. Clearly, a real model of reheating is called for. The assumed decays of the field into fermions rather than bosons is rather slow.

A possible alternative is that the field first decays into bosons, which can happen very rapidly, giving a rapid reheating. The bosons then decay and interact, thermalizing at a high energy. The physics of this reheating clearly is not understood. That being the case, a more general scheme is to assume reheating with a parametrized equation of state and to ask for a full thermal and scale factor history. The main experimental constraints are a data point for the CMB power, mentioned already as a normalization for fluctuations, and spectral index, n_s. The number of e-folds for reheating is N_R with a final reheating temperature of T_R. It is shown later that the data on CMB power require that $V = (0.0054 M_p)^4$. The number of e-folds during inflation, the slow roll parameters, and the ratio of tensor to scalar power, r, can be found for a quadratic power law potential, $V = m^2 \phi^2 / 2$, all of which were previously quoted:

$$\varepsilon = \delta = 1/4x^2, \quad x = \phi/M_p, \quad n_s = 1 - 4\varepsilon, \quad N_e = 2\pi x^2, \quad \phi(t_e) = M_p/2\sqrt{\pi}$$

$$(6.16)$$

More generally, the power law $V = a^{4-b} \phi^b$ determines $\varepsilon = b/4N_e$, $\delta = (b-1)/2N_e$, $r = 4b/N_e$, $n_s - 1 = (2+b)/2N_e$, and the user can choose other exponents in the potential. After the reheating epoch, the standard HBB evolution occurs, beginning with an RD epoch. The total N from the start of inflation until the present is taken to be $\sim$ zero as is appropriate to scales of cosmological interest as mentioned previously. A constant H for inflation is assumed for simplicity. The relationship of pressure and energy density is defined by the equation of state, $p = \omega\rho$, which means that ω for pressure-less matter is $0, 1/3$ for radiation, and -1 for a cosmological term. The equation of state exponent is chosen by the user for the poorly understood reheating epoch. The field at the end of inflation, when $\varepsilon = 1$, is $\phi_e/M_p = b/(4\sqrt{\pi})$. The resulting full expansion history of the scale factor of the Universe appears for specific choices made by the user. Different numbers of e-folds for inflation, reheat, RD, and MD epochs are shown with the constraint that the sum of all N is near zero. The exit and re-entry horizons are $c/a_e H_e$ for inflation and $\sim e^{N_e}$ for a scale near the present horizon:

$$a_e/a_o = \exp(-N_e - N_R - N_{RD} - N_{MD}),$$
$$N_e = \ln(H_e/H_o) - (N_R + N_{RD} + N_{MD}) \tag{6.17}$$

The number after horizon crossing occurs early in the inflationary epoch and is almost the total number of e-folds, $N_e \sim 57$. The number during pre-heating is 5.2 , followed by an RD N of 44 and then an MD N of 9 for the choice of a quadratic potential.

Reasonable values for the equation of state parameter for reheating might vary from zero (matter) to 1/3 (radiation). In the context of a single field causing inflation and of a power-law form, depending on the value of n_s required by accurate CMB data, there may be tension with a pre-heat temperature rising to the GUT scale or even the nucleogenesis energy scale. A matter-like equation of state leads to the highest pre-heat temperatures. A general equation of state parameter is used to approximate the character of that poorly understood physics.

The scale factor, N, for the full history from the start of inflation until the present, for the specific parameters chosen is shown in Fig. 6.17. The line segments, left to right, are for inflation, reheating, RD, and MD. The user chooses both the power-law exponent and the reheat equation of state exponent. Reheating begins with the decay of the field into GUT (?) particles. These particles then decay into the SM particles which cascade down to the "stable" particles of the SM, neutrinos, photons, electrons, and nucleons. The nucleons are contained, after nucleosynthesis, in hydrogen, deuterium, lithium, and helium. The DM is assumed to be heavy, about 1 TeV, but that is an arbitrarily low limit since no compelling DM candidate is known at present.

```
% power law potentials - single field, CMB power and ns fixed by
data
% use full thermal history to predict reheat Temp and e folds
Mp = 1.2 .*10 .^19;    % GeV - Planck mass
As = 2.20 .*10 .^-9 ;    % CMB power-input data for now
ns = 0.965 ; % CMB data -> spectral index
bb = 2 ;% Power of Field in Potential, quadratic
Ne = (bb+2) ./(2 .*(1-ns)) % number of e-folds
```

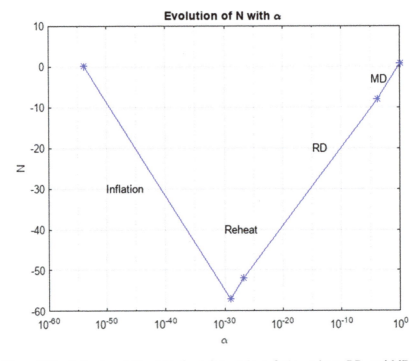

Figure 6.17: Estimates of N e-folds for, left to right, inflation, reheat, RD, and MD as a function of α.

```
Ne = 57.1429
```

```
eps = bb ./(4.0 .*Ne) % slow roll epsilon
```

```
eps = 0.0088
```

```
del = (bb - 1) ./(2.0 .*Ne) % slow roll delta
```

```
del = 0.0088
```

```
r = (4.0 .*bb) ./Ne % slow roll tensor/scaler ratio
```

```
r = 0.1400
```

```
wr = 0.2; % Eq. of State for Preheat = (0,0.333) matter, radiation
gr = 100;   % dof during RD - SMPP
Hk = pi .*sqrt((4.0 .*pi .*As .*(1 - ns)) ./(bb + 2)) % Hk/Mp
```

```
Hk = 4.8862e-05
```

```
Vend = (6.0 .*pi .*pi .*As .*(1-ns) .*bb .*(1-ns)) ./(2 .*(bb+2)) %
Potential at End of Inflation = Vend/Mp^4
```

```
Vend = 3.9898e-11
```

```
% inflation assumed to be exponential in a, linear in N
% preheat phase - from inflation end to preheat
% matter dominated phase, go scaling from present
% rad dom epoch, attach to present temp and reheat
```

The quadratic model yields a short reheating epoch and a large enough reheating temperature. However, the vast range of reheat temperatures for different power laws and different equations of state really indicates that the present "model" is much too crude to offer any predictive power.

Chapter 7

The Cosmic Microwave Background (CMB)

"Cosmology, for centuries consisting of speculation based on a minimum of observational evidence and a maximum of philosophical predilection, became in the twentieth century an observational science, its theories now subject to verification or refutation to a degree previously unimaginable."
— Norris S. Hetherington

"Incidentally, disturbance from cosmic background radiation is something we have all experienced. Tune your television to any channel it doesn't receive, and about 1 percent of the dancing static you see is accounted for by this ancient remnant of the Big Bang. The next time you complain that there is nothing on, remember that you can always watch the birth of the universe."
— Bill Bryson

It was previously argued that with mass scales for the scalar field and the associated potential comparable to M_p, that quantum mechanics will affect the scenario of inflation. In addition, the mass scale means that GR should continue to be the foundation of the analysis since gravity is strong during inflation. However, since a quantum form of GR is not available, a hybrid method needs to be formulated and the resulting predictions examined. Much of the SMC is founded on precise measurement of the CMB, so a more careful analysis is warranted in this section. The CMB has an almost perfect BB spectrum which indicates thermal equilibrium. After decoupling, the CMB still retains a BB spectrum even after it falls out of thermal equilibrium and free streams because the spectrum scales with frequency to maintain the spectrum, albeit dramatically red-shifted, $d\rho_{\mathrm{BB}}/d\omega \sim \omega^3/(e^{\omega/T} - 1)$, $T \sim 1/\alpha$.

There are two distinct fluctuations. Previously, the quantum fluctuations were explored, Eq. (6.11), and the fluctuation, δ_H, with the associated power, P_H, which depends on the slow roll parameter of the potential, V, that drives the effect, ε. This section explores the inflation-created gravitational waves and the density perturbations which create structure in the CMB photons which are observable today. These perturbations are due to metric fluctuations and are labeled by δ_h and P_h since h is the perturbation to the flat space metric — Eq. (7.1).

7.1 Inflation and Gravity Waves, h

A schematic of the assumed resulting history of the Universe is shown in Fig. 7.1. There was a field, necessarily with quantum fluctuations. The perturbed metric led to gravity waves and waves with fluctuations in the field density. These waves were inflated by factors $\sim 10^{26}$ which made them of presently observable size. There followed the end of inflation, a period of reheating which initiated the HBB. Somehow there remained a small excess of matter. As cooling continued, there ended up a plasma of the lowest mass particles of the SMPP, electrons, photons, neutrinos, protons, and neutrons. It is necessary to explore the gravity waves and quantum field fluctuations driven by inflation since they are responsible for the small anisotropies in the CMB and in the related residual structures in matter.

During inflation, there are quantum fluctuations in the field driving inflation. Since the total energy density of the Universe is dominated by the scalar field potential, V, fluctuations in the field, ϕ, lead to fluctuations in the energy density. Due to the rapid expansion of the universe during inflation, these fluctuations in the energy density are frozen outside the causal horizon. Later, in the radiation or matter-dominated era, they will come back within the Hubble radius as they reenter the horizon. As seen in Fig. 7.1, the wavelengths are stretched by the HBB expansion after they re-enter the horizon and can be causally affected. Evolution of the fluctuation amplitudes does not occur when they are outside the horizon; they are "frozen". After inflation, the perturbations re-enter the particle horizon and begin to evolve causally. The density perturbations seed the evolution of the CMB and the large-scale structure of matter. Inflation is indeed a "free lunch" since the detailed structure of the present Universe follows from intrinsic quantum fluctuations and inflation.

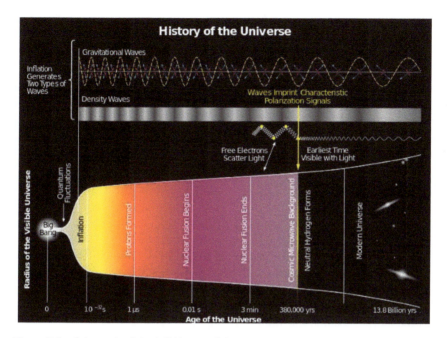

Figure 7.1: Schematic of the full history of the Universe, with emphasis on the two wave types generated during inflation.

The gravitational waves are the simplest and are treated first. The quantum density waves require a full-blown Boltzmann treatment and are treated later. The weak GR metric is the SR metric with added metrical distortions $h_{\mu\nu}$:

$$g_{\mu\nu} = \begin{bmatrix} 1 & 0 & 0 & 0 \\ 0 & -1 & 0 & 0 \\ 0 & 0 & -1 & 0 \\ 0 & 0 & 0 & -1 \end{bmatrix} + h_{\mu\nu} \tag{7.1}$$

The lowest order non-zero elements of $h_{\mu\nu}$ are $h_{oo} = -2\Phi$, $h_{ij} = -2\Phi\delta_{ij}$ (units $\hbar = G = k = 1$ here), where Φ is the weak GR version of the Newtonian potential. The geodesic equations for this metric are then, since the weak field Φ is the Newtonian gravitational potential, the Poisson equation. Newtonian physics is recovered in the weak field limit of GR, as must be the case. The radiative solution for a point object was seen already in the discussion of gravitational radiation, where the metric perturbation is due to the acceleration of the mass quadrupole moment, Q:

$$h = (2/c^4 M_p^2)\frac{d^2}{dt^2}Q(t - r/c), \quad G = \hbar c/M_p^2 \tag{7.2}$$

In the CMB, one expects to see the effects of both tensor and scalar perturbations. There will be gravity waves and scalar density waves due to quantum fluctuations, both of which are inflated to observable sizes during inflation. The gravity waves obey a geodesic wave equation which is damped by the Hubble expansion and sourced by the stress–energy tensor, $T_{\mu\nu}$. The GR Einstein equations for the small metrical perturbations are approximately

$$\nabla^2 h_{\mu\nu} - \frac{d^2}{dct^2}h_{\mu\nu} \sim -(16\pi/c^4)(T_{\mu\nu} - T/2) \tag{7.3}$$

In general, GR is now to be applied to the R–W metric. The linearized equations for tensor perturbations modify the spatial components of the metric with a dimensionless perturbation $\sim a^2(1 + h)$, where h is taken to be small. The linearized Einstein equations for h, where conformal time τ is used as the variable for differentiation, are

$$\frac{d^2}{d\tau^2}h + \left(2\frac{da}{d\tau}/a\right)\frac{d}{d\tau}h + k^2 h = 0 \tag{7.4}$$

Note that k is a dimensionless wave vector here and that a Fourier analysis is made to simplify the differential equations by converting them into algebraic equations. The solutions are gravity waves, that is, neglecting Hubble damping $h \sim e^{ik\tau}$ or waves traveling at the speed of light. The solutions are found symbolically in the following. They induce perturbations in the fluids of the CMB relating metrical perturbation to matter and radiation perturbations as imprinted on the CMB. These decouple at $\tau/\tau_o \sim 0.023$, $\tau_{dec} = 0.076$ when the charged CMB plasma becomes neutral hydrogen and the photons free stream. These perturbations evolve independently of the quantum scalar perturbations, Ψ, Φ to be explored later. The solutions are basically those of a Hubble-damped harmonic oscillator. As such, they have analytic solutions for RD or MD.

The conformal time, τ, is the comoving horizon and the perturbation h is initially $\sim$ constant. For $k\tau < 1$, the perturbation is frozen outside the horizon, while for $k\tau > 1$, the horizon has been re-entered post inflation and causal contact is possible. It will be convenient later to work in

Fourier space in most cases. The wave vector k is a dimensionless conformal wave vector. For $k\tau > 1$, the wave is rapidly, causally, damped by the RD expansion. Large $k\tau_o$, small scale λ, exit late and re-enter and decay earlier, as already seen in Fig. 6.14. This implies that a search for the imprints of gravity waves in the CMB should concentrate on the unique "B mode" detections at large angular scales, Λ, or small modes K. This as yet undetected mode is a key prediction of inflation and needs to be verified.

The analytic solution of Eq. (7.4), with spherical Bessel function solutions, in an RD epoch is shown below in Fig. 7.2. Smaller $k\tau_o$ scales exit the horizon earlier and re-enter later.

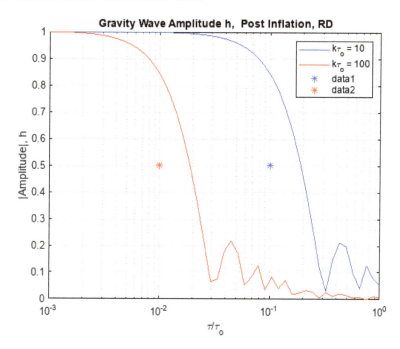

Figure 7.2: Amplitude of gravity waves, h, normalized to 1 at early times, for two different k values as a function of conformal time. The dots indicate horizon crossing.

```
% Find gravity wave amplitude with Hubble expansion
% sub and super horizon boundary, ktauo ~ 1
syms  b y(t) k t ysol
% Initialize   - get Differential Eq to solve
% D2y+(Dy*b)/t +k*k*y = 0 ; b = 2aH using tau.
```

```
bb = 2;
% Initial/Boundary Condition y(0)=1, Dy(0)= 0, symbolic solution
Dy = diff(y);
D2y = diff(y,2);
ysol = dsolve(D2y == -k*k*y -(b*Dy)/t);  %y(0) == 1); ,Dy(0)== 0);
simplify(ysol)
```

$$\text{ans} = -\frac{t^{\frac{1}{2}-\frac{b}{2}}\left(C_2 \mathrm{J}_{\frac{b}{2}-\frac{1}{2}}(kt) - \left(C_1 \cos\left(\frac{\pi b}{2}\right) + C_2 \sin\left(\frac{\pi b}{2}\right)\right) \mathrm{J}_{\frac{1}{2}-\frac{b}{2}}(kt)\right)}{\cos\left(\frac{\pi b}{2}\right)}$$

```
% Solutions are Bessel functions, initial conditions
C1 = 0; C2 = 1; b = bb;
```

Gravitational waves create a density fluctuation amplitude proportional to H during inflation with a power proportional to H, $\delta_h \sim (\hbar H/M_p)$. In the quadratic model, $\delta_h \sim (m/M_p)(\phi(0)/M_p)$. There is a dimensionless fractional amplitude, δ_h, for such fluctuations which is analogous to the scalar amplitude δ_H and can easily be evaluated in the quadratic scalar model. The tensor amplitude is directly proportional to H and a measurement of the tensor fluctuations would determine the energy scale of inflation independent of any issues about the shape of the potential. The ratio of δ_h to δ_H contains the dimensionless slow roll parameter ε.

The gravity perturbations may only be observable in the CMB first at large angular regions with angular index $\ell < 100$ because they die off soon after re-entering the horizon. Only modes with $k\tau_o < 100$ persist. Observable gravity waves in the CMB are first to be seen on large angular scales. The power spectrum, assuming Gaussian fluctuations, with v the velocity field of the fluctuation, is

$$P_h(k) = 16\pi\hbar c/M_p^2([v(k,\tau)/a]^2) \tag{7.5}$$

The gravity perturbation has both a scale, a, and a velocity, v, which appear in the Boltzmann equation. The differential equations for a and v, in Fourier space, are

$$\frac{\mathrm{d}^2}{\mathrm{d}\tau^2}a = 2a/\tau^2, \quad \frac{\mathrm{d}^2}{\mathrm{d}\tau^2}v + (k^2 - 2/\tau^2) = 0 \tag{7.6}$$

Note that the solutions are now in Fourier space, not the ordinary space. Since k and τ appear in Eq. (7.6), it is, one hopes, clear that the equation and its solution are in Fourier space. They can be solved for a and v and the solutions define a power carried by the wave: If H were constant during inflation for all the modes of interest, then the power spectrum would be universal and scale invariant. The fluctuations due to gravity are labeled by δ_h, with $P_h \sim \delta_h^2$. There is no slow roll factor for the gravitational power. The power is defined such that $k^3 P_h(k)$ is dimensionless. The power is defined to be a 2-point correlation function in k space, normalized to a 3-d k volume. The factor $(\hbar/c^2)^2$ which makes the power times k^3 dimensionless is omitted for simplicity:

$$k^3 P_h(k) \sim 8\pi H^2/(M_p^2)|_{k=aH/c} \qquad (7.7)$$

After the k mode leaves the horizon, at $k = aH/c$, $\tau = c/aH$, the power in a mode as defined here is constant and is approximately the same for all modes scaled by $/k^3$. This "scale-free" power spectrum is true if H is $\sim$ constant during all horizon crossings of the different relevant k modes. If these waves are detected, it is seen that the energy scale, H, during inflation would be measured by finding the power in the waves. This would be a new and fundamental constraint on inflation models.

The gravity waves also have a spin as does the photon, in this case, a spin $= 2$ (tensor) units compared to 1 (vector). There are many possible polarization modes. A cartoon of the modes which are irrotational, E modes, and with rotation, B modes, is shown in Fig 7.3. Observation of B modes in the CMB data would be the "smoking gun" of inflation, but alas they have yet to be observed. However, new experiments are being mounted, explicitly with the aim of detecting such modes.

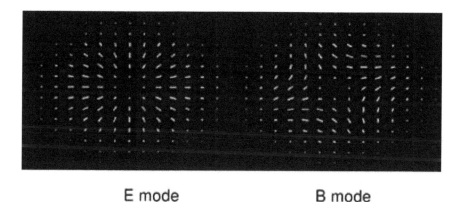

E mode   B mode

Figure 7.3: Schematic of gravity wave polarizations. The B mode is unique to tensor gravity waves.

Figure 7.4 illustrates the experimental difficulties. The surviving gravitational modes due to inflation are at very large angular scales Λ, which means only a few patches of the sky are available, which limits the statistics. There are also backgrounds caused by polarizations induced by foreground lensing or intervening dust scattering which limits the lower bound on the slow roll r parameter, which is the indication of gravitational waves. The BK (BICEP-Keck) data have made a de-lensing correction. The interstellar dust limits the search frequency range with an optimal frequency $\sim 100\,\text{GHz}$. New experiments are being constructed to address these issues and improve the search limits for gravity waves. At present, the constraints on gravity waves are not very restrictive, just an upper limit on the r parameter. Since a Fourier mode must be within the horizon when the CMB decoupling defines the CMB structure, $\tau_{\text{dec}}/\tau_o \sim 0.02$, only h modes with $k\tau_o < 50$ are expected to be relevant. The photons do not fully free stream. They may scatter off the intervening dust in passage through the foregrounds.

It is notable that an August panel has defined, in Dec., 2023, the search for gravity wave signatures available at low multipoles to be a major priority for particle physics research in the next decade. In addition, it endorsed a further extension of the WIMP searches to get to the neutrino "floor" and push somewhat beyond it if possible. Cosmology-related searches are given high priority.

7.2   Scalar Field Inflation: $\phi, \delta\phi$

In addition to tensor modes due to metrical perturbations causing gravity waves, there are scalar perturbations due to quantum fluctuations of the field resulting in density waves (Fig 7.1). Therefore, inflation predicts the structures within the visible CMB. Indeed, the inflation paradigm is what makes the CMB predictions quantitative and then accurately defines the parameters of the SMC. Since the total energy density of the Universe is assumed to be dominated by the inflation potential, V, fluctuations in the inflation field, ϕ, lead to fluctuations in the energy density. Due to the rapid expansion of the Universe during inflation, these fluctuations are also "frozen" into super-Hubble size perturbations. Later, in the radiation or matter-dominated epoch, they will come back into the horizon, as mentioned previously (Section 6.8).

A schematic of the predicted strength of both scalar and tensor perturbations is shown in Fig. 7.5 for the CMB. The dominant scalar effect,

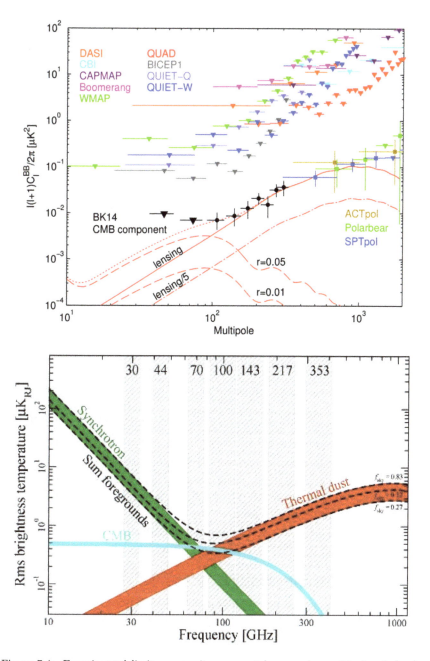

Figure 7.4: Experimental limits on gravity waves at low angular multipoles, ℓ, (top). Foreground and dust backgrounds vs. the CMB signal as a function of frequency of observation (bottom).

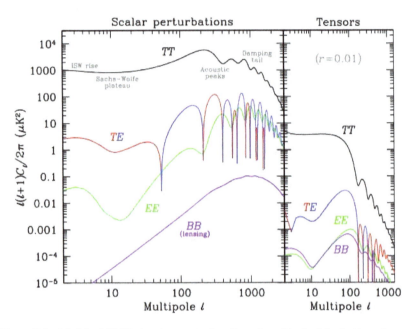

Figure 7.5: Model of CMB structure as a function of multipole ℓ for both scalar and tensor perturbations.

which largely defines the CMB acoustic peaks and other structures, is called TT. The acoustic oscillation peaks occur for $\ell > 100$. For the scalars, the TE and EE modes are subdominant, while the BB mode is induced by lensing and is not primordial but constitutes a background to tensor TT modes. The gravitational tensor mode TT is largest at large angular scales as expected, due to the rapid decay after horizon crossing, Fig. 7.1. The size depends on the r slow roll parameter, shown here for $r = 0.01$, which is approximately the current limit.

A generic scalar field, ϕ, has quantum fluctuations, $\delta\phi \sim \hbar H$. The quantum nature of the fluctuation is evident once the $\hbar$ is explicitly inserted. In this text, units with $\hbar = 1$ will often be used in the interest of simplicity. However, as mentioned previously, judicious applications of factors of c and $\hbar$ will restore equations to the proper dimensionality. The field ϕ has a mass/energy density and pressure, due to the potential, and the time derivative of the field as source terms:

$$\rho = (1/2)\left(\frac{\mathrm{d}}{\mathrm{d}t}\phi\right)^2 + V(\phi), \quad p = (1/2)\left(\frac{\mathrm{d}}{\mathrm{d}t}\phi\right)^2 - V(\phi) \qquad (7.8)$$

The mass density is the GR/QM analog of the classical total energy being the sum of kinetic and potential energy. The pressure is also an energy density source in GR. The equations of motion for ϕ in terms of coordinate time, t, and conformal time, τ, are

$$\frac{\mathrm{d}^2}{\mathrm{d}t^2}\phi + 3H\frac{\mathrm{d}}{\mathrm{d}t}\phi + \frac{\mathrm{d}}{\mathrm{d}\phi}V, \quad \frac{\mathrm{d}^2}{\mathrm{d}\tau^2}\phi + 2(\mathrm{d}a/\mathrm{d}\tau)\frac{\mathrm{d}}{\mathrm{d}\tau}\phi + a^2\frac{\mathrm{d}}{\mathrm{d}\phi}V \qquad (7.9)$$

There appears the familiar "Hubble friction" term and a driving "force" due to the time dependence of the potential. It will not be explicitly stated whether a dynamic equation is in Fourier or normal space. However, it should always be clear in context: t for normal, τ for Fourier; $\frac{\mathrm{d}}{\mathrm{d}x}$ for normal, k for Fourier (for example, Eq. (7.10)). The "friction" due to Hubble expansion damps the field which oscillates at frequencies set by the derivative of the field potential V. A generic scalar field, ϕ, during inflation will quantum fluctuate about the mean with $\delta\phi = \phi - \langle\phi\rangle$. The quantum fluctuation itself, $\delta\phi$, behaves formally like a damped harmonic oscillator during inflation. This then has the same basic form as the tensor perturbation h and the solution has approximately the same behavior. The field equation for fluctuations, in Fourier space, using conformal time as the variable, with a dimensionless wave vector k, is

$$\frac{\mathrm{d}^2}{\mathrm{d}\tau^2}\delta\phi + 2a\mathrm{H}\frac{\mathrm{d}}{\mathrm{d}\tau}\delta\phi + k^2\delta\phi = 0 \qquad (7.10)$$

A dimensionless variation is $\delta \sim \delta\phi/M_p$. It is the fluctuations in the field that are important since they will be imprinted on the CMB. The power in the gravity waves, assuming simple Gaussian fluctuations, was shown previously, $P_h = (8\pi/k^3)(H/M_p)^2$, evaluated at horizon crossing for this mode, $aH/c = k$. For the scalar perturbations, the Fourier analysis of the correlated fluctuations is the dimensionless quantity:

$$\Delta^2(k) = (2k^3/\pi)\int_o^\infty \langle\delta(x)\delta(x+r)\rangle(\sin(kr)/kr)(r^2\mathrm{d}r)$$

$$\sim 8V/(3M_p^4\varepsilon) \quad \text{for } k = aH/c \qquad (7.11)$$

In this case, because the potential is decreasing, the power depends on the slow roll parameter, ε, reflecting the underlying dependence of the field on $\frac{\mathrm{d}}{\mathrm{d}\phi}V$. Compared to the gravity waves, it is expected that the scalar field is dominant since the slow roll parameter ε is small and appears in the denominator of the power. It is simplest to always assume Gaussian fluctuations in the field unless evidence is found for non-Gaussian perturbations.

The quantum fluctuations of ϕ are then inflated outside the horizon. Later, they re-enter, greatly enlarged, and induce structure in the CMB and matter distributions as did the tensor fluctuations.

In general, in GR there are two perturbations in the R–W metric, $g_{oo} = -(1+2\Phi)$ and $g_{ij} = a^2 \delta_{ij}(1+2\Psi)$. A weak field perturbation with isotropic stress is assumed here for simplicity, $\Phi = -\Psi$, and only the potential Φ is retained in the interest of simplicity. The power in the field Φ after exiting the horizon is what will henceforth be tracked as is conventional because the post-inflation power of Φ is simply related to the spectrum of $\delta\phi$ at horizon crossing. The power in the scalar field, ϕ, the potential, Φ, and in the gravitational metric perturbation, h, are shown in the following. The potential is a factor in the CMB because it interacts with the matter components of the CMB:

$$k^3 P_{\delta\phi}(k) = \left(\frac{H}{M_p}\right)^2 \Big/ 2, \quad k^3 P_\Phi(k) = (8\pi/9)(H/M_p)^2/\varepsilon|_{k=aH},$$

$$k^3 P_h = 8\pi/(H/M_p)^2\}_{k=aH} \tag{7.12}$$

These expressions apply to modes which are within the horizon. There was a lot of formalism needed to describe the behavior of h, $\delta\phi$, and Φ both during and after inflation. Unfortunately, it is necessary to quantitatively describe the CMB and matter structures. However, after the appropriate Boltzmann equations are approximated, the CMB structures can be predicted and the determination of the SMC parameters can be understood and appreciated. The treatment here is simplified but not so simple as to lose the physics of the CMB.

7.3   GR Potential, Super Horizon

The idea is to formulate a linearized GR for small inhomogeneity about a background R–W metric in order to study the effects of the small perturbations. The energy/mass fluctuations, because mass defines the metric, lead to a perturbed R–W metric which in turn induces a perturbation in the density of radiation and matter. A linear expansion of temperature (photons), the DM number density, and the "baryon" (e, p) number density is to be made. The result is a set of coupled differential equations, the Boltzmann equations, similar to those encountered in nucleosynthesis, and the WIMP search discussions. In order to understand the details of the CMB, all the couplings between the constituents need to be defined and tracked. In particular, the potential must be one of the sources, as it is for NR Newtonian physics.

The linearized metric with the dimensionless potentials Ψ, Φ is

$$d^2s = -(1 + 2\Phi)(c\mathrm{d}t)^2 - a^2(1 + 2\Psi)(d^2r + r^2d^2\Omega) \qquad (7.13)$$

In the weak field limit, the geodesic field equations approach the New-tonian Poisson equation for the potential Φ sourced by the mass density ρ. In the case of weak GR, the potentials are the same, $\Phi = -\Psi$. Note that there is a spatial curvature perturbation now induced in the zero-order flat R–W space. In the limit that the perturbation wavelength is much less than the horizon, $k\tau << 1$, the potential satisfies the Poisson equation.

It is conventional to make a Fourier analysis in order to convert temporal differential equations into algebraic equations. The conventional variable is k, the dimensionless comoving wave vector. Conformal time, $d\tau = c\mathrm{d}t/a$, is used to track the time evolution of the systems. Horizon crossing occurs first when the Fourier mode, k, is driven outside the horizon by inflation, $k\tau < 1$. Then, after the end of inflation, the CMB relevant mode re-crosses the horizon as the scale increases in an RD and then MD epoch. $k\tau = 1$, $k \sim aH/c$. A physical wave vector presently re-crossing the horizon has $K_o \sim H_o/c$, $K_o = 0.00022\,\mathrm{Mpc}^{-1}$, for example. The wavelength λ outside the horizon is frozen, at early times. However, it will ultimately re-enter. For early times, $k\tau < 1$ outside the horizon Φ is frozen as a constant. Large scales enter later in the MD epoch and smaller scales in RD. The smaller scales, for example, $K \sim 10\,\mathrm{Mpc}^{-1}$, are relatively suppressed due to being affected by the RD radiation pressure.

The behavior of the potential is shown in Fig. 7.6 for modes already looked at for gravitational perturbations. The expression plotted is an approximate one. The large-scale modes of the potential decrease some-what after RD/MD equality at a_{eq} but only by a small amount, $\Phi(\tau)/\Phi(0) \sim 1$, for small K. Large-scale potentials, outside the horizon at decoupling, can be approximated as a constant. Smaller scale poten-tials will experience large reductions upon re-entry to the horizon, similar to what was shown previously for the gravitational perturbation, h. The potential, Φ, post inflation, will be tracked to explore its effect on CMB formation. The plot is for a mode with $K = 0.001\,\mathrm{Mpc}^{-1}$. For comparison, the present horizon occurs at $K_o = 0.00025\,\mathrm{Mpc}^{-1}$.

```
% Find behavior of potential outside and crossing horizon, large
scale
% approximately constant outside horizon;
% variable is y = a/aeq
aeq = 1.47 .*10 .^-4;
zeq = 6780; % ignoring v
% Potential Crossing a_e_q
```

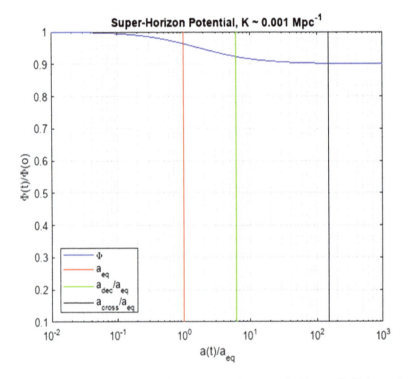

Figure 7.6:   Potential is $\sim$ constant for $k\tau < 1$ as a function of $a(t)$ near CMB decoupling time.

The metric potential will act on the dark matter fluctuations with the potential variations inducing proportional DM fluctuations. The potential is $\sim$ constant (frozen) while outside the horizon for small scales. If it re-enters during the RD epoch, it begins to rapidly decay much like the gravity waves. The equation for the τ development of the potential describes a damped oscillator, with solutions again being Bessel functions:

$$\frac{\mathrm{d}^2}{\mathrm{d}\tau^2}\Phi + (4/\tau)\frac{\mathrm{d}}{\mathrm{d}\tau}\Phi + k^2\Phi/3 \tag{7.14}$$

The solutions are damped as usual by a Hubble-like term, as seen in Fig. 7.7. For $k\tau = 10$, the falloff is large for $\tau > 0.1$, soon after τ_{eq}. Re-entry well before the MD epoch means that the potential decays rapidly. However, modes do persist until $\tau_{dec} = 0.076$ during the time of CMB structure formation.

```
% find potentials in RD epoch
% diff eq for phi:  d2y = -4dy/t - k^2y/3 = 0, tau variable
% solution is Bessel function - as done before, Jv(z)
```

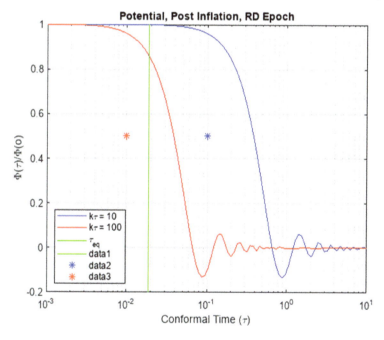

Figure 7.7: Potential as a function of τ for two different k values. Horizon crossings are indicated by the stars.

7.4 Horizon Cartoon

This script generates a cartoon, shown in Fig. 7.8, of the potential $\Phi(\tau)$ during constant H inflation, starting outside the horizon and later re-entering.

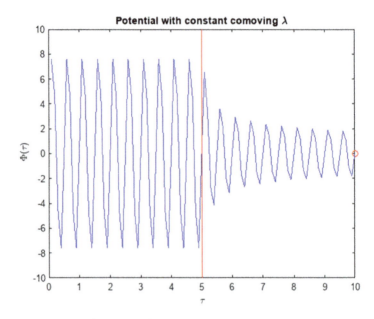

Figure 7.8:   Cartoon of a Fourier k mode both outside the horizon and crossing it at $\tau = 5$.

A related amplitude appears in Fig. 7.7 and Eq. (7.14) shows that the potential is oscillatory with a damping which is large during an RD epoch. This is a simple and schematic cartoon, but it may make the horizon crossing a bit more graphic. Time runs from sub to super horizon, as the mode exits the horizon. In sub-horizon, there is a constant comoving wave vector, k, and amplitude since the mode is frozen outside any causal contact. After horizon crossing, the physical wavelength increases with expansion and the amplitude decreases, but the comoving wavelength is constant. The boundary is chosen to be a fixed conformal time, $\tau = 5$, crossing the horizon and damping with $k = 1/5$.

```
% Look at potential fluctuations in ~ constant H inflation
% pick an oscillation, Movie for mode
% DH = c/H, constant H during inflation
% horizon crossing at tau = 5, ktau = 1 k = 1/5, conformal wave
vector
```

7.5 Boltzmann Equations: Perturbations and Potentials

The increasingly precise CMB data have largely driven the determination of the parameters of the SMC. Although there has been a fair amount of formalism, nevertheless it is important for the reader to see how those data are such that the parameters can be extracted without "degeneracy", that is, multiple solutions with different parameters yielding the same CMB predictions within experimental errors. It is necessary to construct a series of Boltzmann equations for all the components contributing to the CMB in order to obtain a better picture of the CMB formation. The treatment will still be simplified but not so simplified as to lose the main feature of the largest scale CMS acoustic structures.

A schematic of the multiple actors in the CMB appears in Fig. 7.9. There are gravitational metric perturbations, h, as well as quantum field fluctuations, Φ, Ψ, and thermal fluctuations, Θ. The dark matter is the simplest since it is assumed to only react with the metric fluctuations. The neutrinos, being weakly interacting and UR at most times, are largely ignored in this text even though the number of primordial neutrinos is

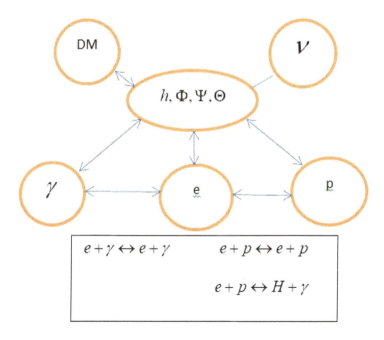

Figure 7.9: Schematic of the interacting components of the CMB.

comparable to that of the photons. The photons, electrons, and protons all react with the perturbations but also with each other through Compton/Thomson scattering and Coulomb e-p scattering. Indeed, it is the polarization induced by the scattering which is a background to the search for the "B mode". At the end of the CMB formation, the photons will decouple from charged particles (baryons so-called) and the formation of neutral hydrogen dominates. At that point, the Universe is transparent to the CMB and optical observations become possible. That happens at an HBB time of ~400,000 years.

A simplified set of Boltzmann equations is used in this text. Neutrinos are largely ignored in the CMB. The gravitational fluctuations, h, are largely relegated to very large scales and have yet to be observed, so they are also ignored. The field potentials are reduced to a single potential, Φ, in the weak GR approximation. A complete analysis is made in actual data fitting and the results are plotted along with the CMB data. However, the simplified model retains the most salient feature of the CMB acoustic structures. The slow roll model for inflation solves the basic problems of HBB cosmology. In addition, gravitational waves are predicted to be presently observable, and quantum fluctuations of the scalar field predict faint present-day structures in the CMB. In order to understand these predictions, a set of coupled Boltzmann equations is needed to describe the elements of the SMC and their evolution. In the MATLAB script, a simplified set of equations is used.

A schematic of a Boltzmann equation for a number density, n, and velocity, v, subject to a process with a cross-section σ is with dimension n/t and a reaction rate $\Gamma \sim n\sigma v$. These equations describe the evolution of the different constituents of a system, as is familiar from the prior discussions of WIMPs and nucleosyntheses:

$$(1/a^3)\frac{\mathrm{d}}{\mathrm{d}t}(na^3) \sim n^2\sigma v \tag{7.15}$$

Quantum perturbations are described by the potential Φ which is the Newtonian potential ϕ in weak field GR. There are fractional density perturbations caused by the metric potentials: $\delta = d\rho/\rho$. There is an equation of continuity and, for momentum conservation, generically. The time development of the perturbation is due to the flow and the Hubble damping and the potential Φ acts as a source. The velocity-time dependence is driven by the gradient of the potential, also with Hubble damping. The perturbation is defined by a fluctuation δ and a velocity v, which are sourced by the space and time dependence of the potential:

$$\frac{\partial^2}{\partial^2 t}\delta + (c/a)\frac{\partial}{\partial x}v + (3/\hbar)\frac{\partial \Phi}{\partial t} = 0, \quad \frac{\partial}{\partial t}v + Hv + (1/a)\overrightarrow{\nabla}\Phi = 0 \quad (7.16)$$

These differential equations become algebraic equations when working in Fourier space, where k is the comoving wave vector, which is therefore much more convenient. Early times mean that perturbations are causally connected and scalar perturbations to the metric couple to matter and radiation. The quantum perturbations which are produced at causal scales and are pushed outside the horizon re-enter much later and are seeds of the CMB structures. The two equations which express the DM perturbation continuity and momentum conservation, where H is here the Fourier version, $H = (d\alpha/d\tau)/\alpha$, are:

$$\frac{d^2}{d^2 t}\delta + ikv = -3\frac{d\Phi}{dt}, \quad \frac{d}{dt}v + Hv = -ik\Phi \quad (7.17)$$

For the CMB, the constituents are the matter: electrons, protons, and neutrinos, plus the gravity waves and dark matter. The dark matter is the simplest since it only gravitates. The dark matter number density fluctuates fractionally by δ_{dm}. It is described by the density fluctuations and a velocity field, δ_{dm}, v_{dm}, where weak field GR is assumed. Note that the fluctuations are now Fourier-transformed spatial fluctuations. Perturbations in the GR field lead to changes in the baryon perturbation, which in turn couple back to changes in the potential. The DM perturbations grow after horizon re-entry approximately as $\delta_{dm} \sim \ln(0.6k\tau)$ driven by the causally active metric potential Φ.

The baryons (electrons and protons — this is a convention, meaning ordinary matter) are tightly coupled by Coulomb scattering, $e^- + p \longleftrightarrow e^- + p$, which means they are describable by a single "baryon" fluctuation, $\delta_e = \delta_b$, $v_e = v_p$. Since the reaction rate $\Gamma_{ep} >> H$, the baryons are in thermal equilibrium. They obey the same equations as the dark matter except for the addition of sourcing by photons which Compton/Thomson scatter off the baryons and add a term to the baryon velocity equation. The full set of equations is deferred until later when the acoustic structures in the CMB will be simulated. One lesson of interest is that the Compton scattering carries polarization, which leads to a prediction that the CMB photons should be polarized. This is a secure prediction of the SMC. The polarization mode, EE or BB, yields information about the gravitational inflation.

There is also a perturbed Bose-Einstein energy distribution function, f, for photons characterized by Θ, $kT \to kT(1+\Theta)$, $\delta T/T \sim \Theta$ which approximately describes a non-zero chemical potential indicating out of thermal equilibrium behavior. For the CMB, the perturbations are expanded in terms of spherical harmonics and the correlations C_ℓ are summed over m. The density perturbations are sourced by both Θ and Φ as seen in Fig 7.9:

$$f \sim 1/[e^{E/kT(1+\Theta)} - 1] \tag{7.18}$$

Sources of dark matter, baryons, and photon perturbation drive the GR potential, Φ:

$$k^2\Phi + 3\left(\frac{\mathrm{d}}{\mathrm{d}\tau}a/a\right)\left(\frac{\mathrm{d}\Phi}{\mathrm{d}\tau} - \left(\frac{\mathrm{d}}{\mathrm{d}\tau}a/a\right)\Phi\right)$$
$$= 4\pi(a/M_p)^2[\rho_{\mathrm{dm}}\delta_{\mathrm{dm}} + \rho_b\delta_b + 4\rho_\gamma\Theta] \tag{7.19}$$

The overall result is a set of coupled Boltzmann equations which depend on the parameters of the SMC and which can be confronted with the measured structures of the CMB. As seen before, Sec. 7.1, for gravity waves, constant H, all modes have the same "scale free" power at the lowest order. For the scalar field, Sec. 7.2, the perturbation is also approximately scale-free but with a slow roll denominator, ε, which means that it is expected to dominate over the gravity wave power. In turn, this means the tensor power is a factor $\sim 9\varepsilon$ smaller than the scaler power. For this reason, in the simplified CMB treatment which follows, the tensor perturbations are ignored:

$$P_h(k) = 8\pi(H/M_p)^2/k^3, \quad P_H(k) = 8\pi(H/M_p)^2/(9k^3\varepsilon) \tag{7.20}$$

7.6 CMB Acoustic Oscillations

It is instructive to first look at the situation when wavelengths are less than D_H and apply Newtonian physics. The Poisson equation relates density fluctuations, $\delta\rho$, to potential fluctuations, $\delta\Phi$, as $\nabla^2(\delta\Phi) = 4\pi\delta\rho/M_p^2$. For the fluid flow of matter, there are pressure-driven acoustic oscillation terms and potential terms. The oscillatory modes have frequencies $\omega = \sqrt{(Kc_s)^2 - 4\pi\rho_o/M_p^2}$ so that on small scales there are sound waves, while on large scales, there is an exponential collapse (Section 8, Jeans Length).

The CMB temperature perturbations occur because photons lose energy climbing out of the gravitational potential, Φ, of the higher-density regions of DM. The measured CMB fractional temperature anisotropies are about a part in 10^5. Specifically, the measured scalar power is $P_H = 2.2 \times 10^{-9}$ so that $V/\varepsilon = 8.2 \times 10^{-10} M_p^4 = (0.0054)^4 M_p^4$. In the previous numerical example for a quadratic potential, $\phi(o)/M_p = 3.25$, and the required value for m was $1.1 \times 10^{-6} M_p$. Indeed, this value for m was previously chosen in a seemingly arbitrary fashion in order to agree with the measured CMB scalar power. With m fixed, the spectral index can be solved for. The quadratic model is fixed by the required number of e-folds which defines the initial field and by the mass which is fixed by the measured CMB power.

The two-point temperature correlation multipole ℓ is $C_{\ell\ell}$. It is the average temperature fluctuation for two points separated by a spherical angle θ, averaged over all such points. The expansion in Legendre polynomials means that an angle θ and a parameter ℓ are related as $\theta \sim \pi/\ell$. Since $D_H(z_{\text{dec}}) \sim 0.2$ Mpc and $z_{\text{dec}} \sim 1100$, the angular distance of the horizon at decoupling is $D_A(z_{\text{dec}}) \sim 3.35(4439) \text{Mpc}/1100 \sim 13\,\text{Mpc}$. Therefore, the horizon at decoupling has a present angular size of about 0.2 Mpc/13 Mpc or about one degree. These approximate values set the stage for a more quantitative analysis.

We are finally able to look at the CMB structures as defined in the IBB model, albeit in a simplified approximation. The CMB data can be well fit to a rather simple set of hypotheses: the SMC. Here, a simplified set of Boltzmann equations is used, which works well in the angular region of the first two acoustic peaks. The aim is to keep it simple but not so simple as to lose the physics. There is only a single potential, Φ, in the weak field GR limit. There is a perturbation in the density and a velocity field for matter and for photons. They are strongly coupled together by Compton/Thomson scattering. In the simplified model, there are only those two fluids — the dark matter and the "baryon"-photon plasma. They obey a set of five coupled equations in Fourier space, two for the dark matter, position fluctuation and velocity perturbations, δ_{dm}, v_{dm}, and two for the analogous photon distributions, δ_γ, v_γ. There is a final equation for the potential itself, Φ. With tight coupling, the "baryons", or ordinary matter, follow the photons, $\delta_b = (3/4)\delta_\gamma$, $v_b = v_\gamma$, which means two fewer variables. All variables are defined, with comments, in the MATLAB script which solves the set of five equations numerically using the MATLAB utility "ode45" as is typical in this text. The first equation is the Fourier space equation for DM which has the simplest behavior, sourced by the potential Φ. The Boltzmann equations for the photon-baryon plasma and the potential are simply taken as given:

$$\frac{\mathrm{d}}{\mathrm{d}x}\delta_{\mathrm{dm}} = -\kappa v_{\mathrm{dm}} + 3\frac{\mathrm{d}}{\mathrm{d}x}\Phi, \quad \mathrm{d}v_{\mathrm{dm}}/\mathrm{d}x = -\tau v_{\mathrm{dm}} + \kappa\Phi$$

$$\frac{\mathrm{d}}{\mathrm{d}x}\delta_{\gamma} = -(4/3)(\kappa v_{\gamma}) + 4\frac{\mathrm{d}}{\mathrm{d}x}\Phi,$$

$$\frac{\mathrm{d}}{\mathrm{d}x}v_{\gamma} = [-(3/4)v_b\tau v_{\gamma} + (1/4)\kappa\delta_{\gamma}]/(1 + (3/4)y_b) + \kappa\Phi]$$

$$\frac{\mathrm{d}}{\mathrm{d}x}\Phi = -\tau\Phi + (3\tau^2/2\kappa)\frac{[v_{\gamma}(4/3 + y - y_{\mathrm{dm}}) + v_{\mathrm{dm}}y_{\mathrm{dm}}]}{(1 + y)} \quad (7.21)$$

The first two equations are for the DM coupled only to the single metric GR potential, Φ, and x is a scaled time variable, $x = \tau/\tau_{\mathrm{dec}}$, y is $(ax)^2 + 2\alpha x$, and $\kappa = k/\tau_{\mathrm{dec}}$ is a scaled wave vector. The DM density perturbation and velocity field are coupled only to the potential and the time derivative of the potential since DM interacts only with gravity. The photon (and baryon — assuming tight coupling) density perturbation and velocity field are coupled to the potential and time derivative plus additional baryon terms. Finally, the potential time derivative couples to the photon velocity field and the DM and baryon sources, similarly. Note that κ is here just a scaled parameter of the model and not the curvature.

This is a simplified treatment, but it is explicable to the user and it yields some of the most salient aspects of the SMC model. Indeed, in the next section, the user can change the basic parameters of the SMC and see how the CMB structures change.

The script fixes the scale factors for α_{dec} and α_{eq}. The time variable is x and is given a range $(0.001, 1)$ while $\kappa = k/\tau_{\mathrm{dec}}$ covers the range $(0.01, 100)$. The multipole range is $1 = (10, 1000)$. The factor to scale from τ_{dec} to the present τ_o is "tweaked" by about 25% from the expected value as needed due to some of the approximations, such as the "sound horizon" which was made in the script.

The Fourier solutions are then projected onto the spherical harmonics using the spherical Bessel functions evaluated with the MATLAB utility "besselj". These angular variables are those appropriate for full sky surveys. The result is numerically integrated over k to get C_ℓ using the MATLAB utility for numerical integration, "trapz". The potential, the photon density perturbation, and the photon velocity field all contribute. Very approximately, $\ell(\ell+1)C_\ell \sim \delta_H^2$:

$$C_\ell \sim \ell(\ell+1)\int \mathrm{d}k[(\Phi + \delta_\gamma/4)j_\ell(k\tau_o) + v_\gamma \mathrm{d}j_\ell(k\tau_o)/d(k\tau_o)]^2 \quad (7.22)$$

Finally, the resulting function is smoothed using a simple polynomial with the utility "polyfit" and the result is plotted. The Bessel function for l peaks at $k \sim \ell\tau_o$ so that a given scale k roughly corresponds to a multipole ℓ, but many k contribute to a given ℓ. The standard CMB analysis expands in spherical harmonics over the full observable sky. Two regions of the sky separated by an angle θ correspond to a correlation multipole ℓ with $\theta \sim \pi/\ell$. Apparently causally disconnected segments of the sky number ~ 50 or about 4 degrees of subtended angle at present so that the horizon for the CMB corresponds to a multipole of $\ell \sim 50$. which is approximately the location of the first acoustic peak.

The total matter and baryonic fractions are user input. The script then asks the user to pick a $k\tau_{\text{dec}}$ value, $\tau_{\text{dec}} = 0.076$. An example of the DM density perturbation solution δ_{dm} and photon-baryon solution is shown in Fig. 7.10. In that plot, the cold matter only begins to increase for $\tau/\tau_{\text{dec}} \sim 0.0025$. The associated plots for the photon-baryon velocity and the potential as a function of conformal time are shown in Fig. 7.11.

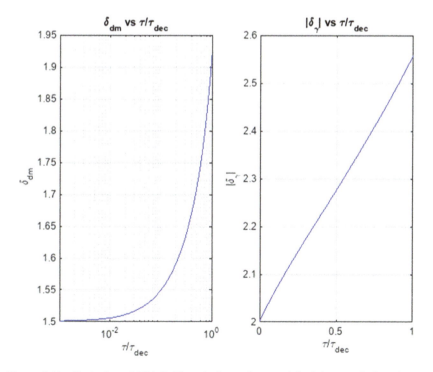

Figure 7.10: Evolution of DM (left) and photon-baryon δ (right) vs. τ before decoupling. Both increase steadily in absolute magnitude.

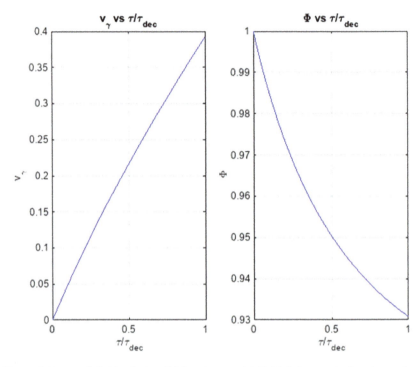

Figure 7.11: $\gamma - b$ fluid velocity (left) and potential Φ (right) vs. τ before decoupling. The fluid velocity increases $\sim$ linearly while the potential is $\sim$ constant.

```
% Numerical solution of a simplified set of equations, Seljak
% CDM - fluctuation and velocity, photon fluid (baryon coupled)
% and velocity plus evolution of metrical potential;
% constants for the system of 5 equations
% 2 for CDM, 2 for photon-b fluid, 1 for the potential and deriv
omm = 0.3; % Total Matter Fraction (SM ~0.31
omb = 0.05; % Baryon Fraction (SM ~0.05)
adec = 1.0 ./1100 ;   % recombination/decoupling scale a
aeq = 2.47e-4 ;% scale at equality with 3 neutrino;
% SM a(t) at equality, decoupling and alpha parameter
% omega rel fixed, compute aeq for this input, assume adec is fixed
% This a(t) at equality, decoupling and alpha parameter
tauo = 3.33;    % numerical integration for SM and 3  v generations
taueq = 0.024;
taudec = 0.068 ;
% approximate damping scale in kappa
 [t,yy] = ode45(@CMB_5_v2,t,[yo1 yo2 yo3 yo4 yo5 ]);
```

The speed of sound in the plasma of photons and baryons is reduced by the baryons since they are heavy, $c_s = \sqrt{1/(3(1+R))}$, $R = 3\rho_b/4\rho_\gamma$. This fact makes the CMB data very sensitive to the baryon fraction using the acoustic peaks. The acoustic peaks are formed at τ_{dec} and observed at τ_o, first the fundamental, with harmonics at higher ℓ values, smaller angular scales, $\ell_{\text{peak}} \sim nK\tau_o \sim n\pi\sqrt{3}(\tau_o/\tau_{\text{dec}}) \sim 239$ for $n = 1$. In the present, quite simple, approximation, the first two acoustic peaks and the valley between them are quite evident.

The observation of a BB spectrum for the CMB shows that it was in thermal equilibrium at decoupling energies $\sim 0.26\,\text{eV}$, $z \sim 1100$, by way of processes like $e^- + p \longleftrightarrow H + \gamma$. The matter is tightly coupled to the photons through Compton scattering, $e + \gamma \longleftrightarrow e + \gamma$, with a mean free path $<<$ the horizon. A cartoon of the acoustic oscillations appears in Fig. 7.12.

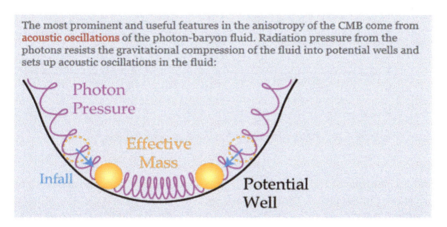

The most prominent and useful features in the anisotropy of the CMB come from acoustic oscillations of the photon-baryon fluid. Radiation pressure from the photons resists the gravitational compression of the fluid into potential wells and sets up acoustic oscillations in the fluid:

Photon Pressure

Effective Mass

Infall

Potential Well

Figure 7.12: Cartoon of the acoustic oscillations of the $\gamma - b$ plasma. Photons lose energy as they climb out of the baryon potential well.

Just as with an harmonic oscillator, the position and velocity are out of phase, as illustrated in Fig. 7.13. At the maximum position, the turning point, the velocity is 0. Since the coupling of the "baryons" to the photons is tight, these oscillations will exist also in the matter and, as will be seen later, appear as baryon acoustic oscillations (BAO) in the matter of the Universe. The higher ℓ acoustic peaks are suppressed because of Thomson scattering, which is treated as a simple damping factor in this simplified treatment. The first peak is large because it is an expansion peak, while the second is a compression peak. Odd-numbered peaks are favored in general

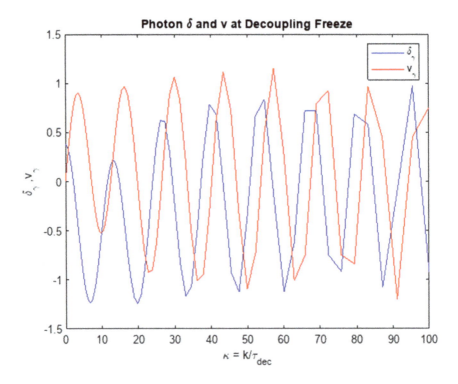

Figure 7.13:   Fourier behavior of the photon potential and velocity.

and an example of the smoothed $C_{\ell\ell}$ is shown in Fig. 7.14. The actual data are shown in Section 7.7 for comparison.

The CMB scaler contributions determine H^2/ε, all modes at horizon crossing, while the tensor contributions would determine H^2 itself. Perturbations at different wavelengths (k) are imprinted by decoupling on the CMB at different phases in their oscillation. Oscillations can appear in the CMB because the Fourier components of the spectrum all coherently exit the horizon at almost the same time which is a distinct prediction of inflation. Since all Fourier modes with a given wavelength have the same phase, they interfere coherently, yielding peaks and valleys in the CMB. Without coherence, the spectrum would be featureless. The creation of the CMB at decoupling/recombination is approximately instantaneous.

The second peak is of smaller amplitude since the baryons oppose this mode which is in a phase of expansion. Higher harmonics at large ℓ or

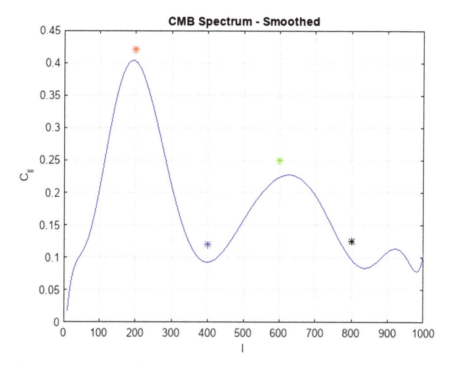

Figure 7.14: Smoothed CMB acoustic oscillation as a function of ℓ showing the fundamental and first three harmonics.

small angles are damped because the fluid is not perfectly coupled, baryon to photon, so that there is an added diffusion term causing weaker oscillations. A typical scale for diffusion is <100 Mpc. The peak height depends on the baryon and DM abundances and helps determine their magnitude since more matter damps the oscillations which are driven by the photons. In this simple model, there is an *ad hoc* damping of C_ℓ with an exponential damping factor with $\ell_d = 600$ because the fluid coupling is tight by assumption in this model. Smaller scales, higher ℓ, are affected most by baryon diffusion.

7.7 CMB and SMC Parameters

There have been a lot of formulae in Section 7. Nevertheless, it is instructive for the user to be able to choose the most important inputs to the CMB angular structures and see how those choices define the agreement of the simplified model to the features of the CMB data. The Planck experimental data are shown for comparison. Note the extreme faintness of the peaks. The scales are tens of μK compared to the mean CMB temperature of 2.726 K^o which is a feat of experimental science.

The Planck collaboration, using data shown below in Fig. 7.15, made an initial fit in 2015 of the CMB data in terms of only six parameters, $\Omega_m, \Omega_b, h, n_s, P_s, L_{\text{ls}}$. The basic fit assumes that the Universe is flat and that the small effects of tensor power can be ignored. The photon number density and temperature are assumed to have negligible error. Neutrino effects are not variable. What remains are the matter and baryon fractions, the Hubble parameter (h), the scalar power and spectral index, and the last scattering length of the plasma.

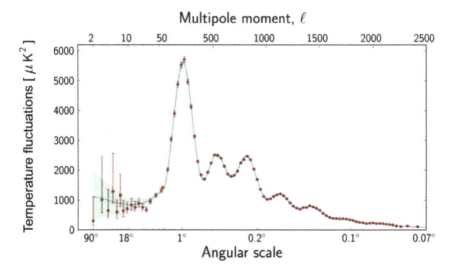

Figure 7.15: CMB data from the Planck collaboration showing several acoustic peaks.

Both the angular scales and the multipole scales are shown. The vertical scale is microKelvin squared μK^2, small with respect to the mean CMB temperature of 2.726 K^o which sets the scale of the perturbations due to inflation. The fundamental mode is at about 1 degree. Higher harmonics occur at integer multiples of the fundamental mode. Small angle has higher ℓ. The lowest multipoles suffer from limited statistics as the samples cover a substantial fraction of the sky. The fundamental peak is at $\Lambda \sim 400$ Mpc. It covers $\sim 1/100$ of the sky, as expected from the discussion of causality with ~ 50 patches of the present sky seemingly casually disconnected. The peak is at $K = 0.016$ Mpc^{-1}.

Most of the SMC parameters can be well determined using the CMB data alone. There is sufficient information that there is no real degeneracy in the parameter space, given the large number of well-determined peaks and valleys arising from the photon-baryon acoustic oscillations. The user chooses two of the main parameters, the baryon Ω and that for all matter, DM and baryons, assuming a flat space. The plots are overlapped so that the user can be convinced that different choices of the parameters yield different and unique shapes for the acoustic oscillations; either the peak heights or location or the depth of the valley varies. The effects of the parameter choices on the height of the first peak and the height and location of the second peak are substantial. Indeed, the actual CMB data are of such high quality at present that the full set of SMC parameters can be quite precisely determined, under certain assumptions. However, a cross-check using the supernova data is to be performed as is a consistency proof.

The following code is largely a repeat of that from Section 7.6 and is included simply to give the user a chance to more easily visualize the changes that occur with changes in two of the key parameters: the total matter and the baryon Ω values. A plot for a specific parameter choice appears in Fig. 7.16. More baryons change the sound speed and hence the peaks, as expected. It should be noted that the higher harmonic peaks are suppressed. This is because there is a residual ionization, as already explored in Figs. 5.10 and 5.11. There is a damping called kd in the following code. The photons do not stream totally freely but Thomson scatter off the residual free electrons. The last scattering length is $L_{\mathrm{sls}} = \sigma T \int_{t_{\mathrm{dec}}}^{t_o} n_e (cdt)$. The scattering washes out the acoustic peaks at smaller scales and higher ℓ values.

```
% constants for the system of 5 equations
% project Fourier onto Ylm with spherical Bessel functions
% integrate over k to get Cll
pol = polyfit(ll,Cll,11);
Ctt = polyval(pol,ll);
```

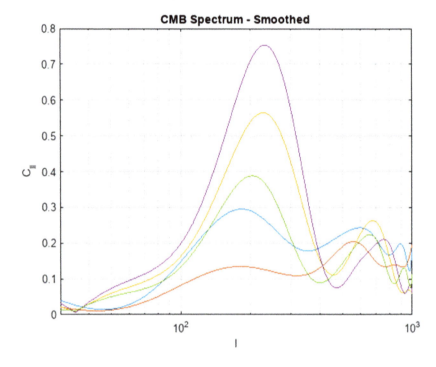

Figure 7.16: Acoustic oscillation spectrum for user choices of SMC parameters.

```
% first acoustic peak - height and location

cpeak = 0.3875
lpeak = 205.4774

ClearPlot =  1 % set = 0 to clear plot, start ove
```

Chapter 8

Large-Scale Structure (LSS)

"The James Webb Space Telescope was specifically designed to
see the first stars and galaxies that were formed in the universe."
— John M. Grunsfeld

"In a spiral galaxy, the ratio of dark-to-light matter is about a
factor of ten. That's probably a good number for the ratio of our
ignorance to knowledge. We're out of kindergarten, but only in
about third grade."

— Vera Rubin

Having followed the formation of the CMB, it is time to follow the tightly
coupled "baryons" and DM to explore stellar formation and the analog of
the CMB photon acoustic oscillations. They are called "baryon" acoustic
oscillations (BAO) and they have the same parentage as the CMB acoustic
peaks and valleys — the tightly coupled photons and "baryons" (electrons
and protons). In fact, the BAO are an essentially independent cross-check
of the determination of the parameters of the SMC. There are two basic
competing effects in cluster formation, gravity and temperature, or thermal
velocity. A basic question for the galaxy surveys was, is the DM hot or cold?
Structure formation data answered this question with the SMC which is
often called the LCDM model: dark energy, cold dark matter, "baryons",
cold neutrinos, and photons.

8.1 Clumping Cartoon

The CMB photons free-streamed after decoupling, preserving the acoustic
structures. The matter also had acoustic footprints, but it subsequently
was very far from free streaming. The CMB description of the photon

structures was accomplished with a model, the SMC, with only a handful of key parameters. With the tight coupling of baryons to photons, at decoupling, the baryons also have an acoustic structure. Observing them is an excellent check on the validity of the model. However, the detected objects are not free-streaming microwave photons but galaxies observed at a variety of z values. Several extensive galactic surveys have been made and new instruments are being constructed and will soon be brought online. However, in no sense has the matter been free streaming, and a rather more complex and nonlinear analysis is needed. This being the case, the ensuing exposition is rather superficial because nonlinear analysis is far outside the scope of this text. However, the observation of the BAO structural peak can be understood by applying simplifying approximations.

First, a very simplified model of how particles would clump together under gravitational attraction is made in the following. Gravity always wins in the end (ignoring DE here). In this case, the clumping would result in the formation of stars and clusters of stars. The user picks the initial number of particles and their velocity (pressure), which is in a random direction. The particles are confined to a 2-d box, but the idea is to get a feeling for how rapidly particles might aggregate. In any case, it is clear that gravity will cause particles to clump together under the gravitational attraction but hindered by pressure. Each particle pair feels an inverse square attraction along the line of centers of all the other particles. If a pair reaches a minimum separation, the pair is merged/clumped. At low velocity, the particles are ~ all merged but with a velocity ~1 the increased temperature/pressure results in reduced clumping. An example of the exercise appears in Fig. 8.1.

```
% look at clumping of a system under gravity - low temp
% select density and "pressure"
% put N particles randomly in a 2-d box and let them
% evolve under mutual gravitational attraction
N = 20 ; % Number of Particles to Track
vo = 1 ; % Velocity (~ 1)
% initial conditions, random locations, start with all with random
velocity
          vx(i) = -vo + 2 .*vo .*rand;
          vy(i) = -vo + 2 .*vo .*rand;
% Initial Mean Separation
```

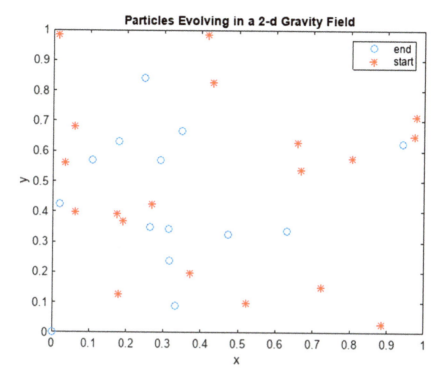

Figure 8.1: Plot of a simplified model of a 2-d system of mass points clumping under gravitational attraction.

```
dRij = 0.0327
```

```
% time steps
dRij = sqrt(dRij) ./(ij);
il % Number of Final Clumps
```

```
il = 14
```

```
dRij %Final Mean Separation
```

```
dRij = 0.0378
```

```
% Movie of Evolution of the System
```

8.2 DM Evolve

Structure formation is largely driven by the dominant DM via gravity. This being the case, the evolution of DM after decoupling is followed. An approximation gives a simple idea of the time of decoupling using the simple $\Gamma = H$ for the transition from thermal equilibrium idea yet again. The reaction rate scales as the number density $\sim 1/\alpha^3$ and the ionized fraction X. The MD Hubble parameter scales as $n/t \sim \alpha^{-3/2}$. Solving for the condition that $\Gamma = H$, a decoupling z of ~ 1100 is found and serves as a cross-check. At decoupling time, the density of DM is about $2 \times 10^9 \, \text{GeV/m}^3$, while baryons have a density of about $3 \times 10^8 \, \text{GeV/m}^3$ and photons are comparable at about $4 \times 10^8 \, \text{GeV/m}^3$. Therefore, the DM dominates cosmic evolution after CMB decoupling. The DM will form the "seeds" into which the baryons collapse and which then define the large-scale structures. The focus in what follows is now on DM for that reason.

In Fig. 7.2, it is seen that gravity waves die off rapidly after horizon crossing so that only large scales are sensitive to tensor perturbations. In Fig. 7.6, it is seen that the potential outside the horizon, $K = 0.001 \, \text{Mpc}^{-1}$, is $\sim$ constant in the RD epoch, but then as seen in Fig 7.7, if a mode is inside the horizon during RD, it decays rapidly. Looking at Eq. (7.17), the DM fluctuation is sourced by the potential, and in Eq. (7.19), the potential itself is partially driven by the DM. The evolution of the matter, DM, δ_{dm}, fluctuations with time depends both on whether the mode is within the horizon and the state of matter or radiation dominance, MD, RD, when the fluctuation can participate in causal physics. For small modes within the horizon, there is growth which is suppressed in RD due to the photon pressure, but then rapid growth resumes in the MD epoch. For a mode outside the horizon, there is rapid growth in the RD epoch slowing during the MD epoch. Growth is stalled if a mode enters the horizon in the RD epoch. For example, $K \sim 10 \, \text{Mpc}^{-1}$ growth in δ_{dm} is only logarithmic as shown below in Table 8.1 and in a cartoon of the different cases in Fig. 8.2. The RD/MD boundary can be taken to occur at $\tau_{\text{eq}} = 0.019$, $\alpha_{\text{eq}} = 1.47 \times 10^{-4}$, $K_{\text{eq}} = 0.0147 \, \text{Mpc}^{-1}$, $\frac{t_{\text{eq}}}{t_o} = 1.3 \times 10^{-6}$. Finally, in the MD epoch, all modes that have crossed the horizon grow with α as illustrated in Fig. 8.4. Note that the analysis assumes linear growth, which is valid only for $K \sim 0.1 \, \text{Mpc}^{-1}$, and nonlinear effects begin to dominate, an effect which is ignored.

Table 8.1: Approximate behavior of the growth of dark matter perturbations δ.

	RD	**MD**
δ inside horizon	$\log(\alpha)$	α
δ outside horizon	α^2	α

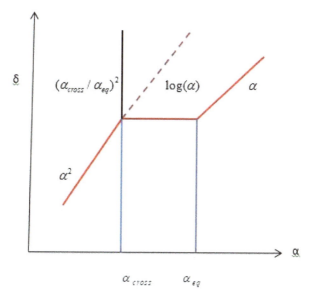

Figure 8.2: Cartoon of the behavior of a DM mode which crosses the horizon in the RD epoch.

The DM perturbation grows with time as it is purely gravitational and not opposed by a photon pressure. In the CMB discussion of Section 7, a slight growth was already seen. The DM Boltzmann equations used in Section 7 can be used, with DM fluctuations sourced by the potential, Φ. In the MD epoch, both matter and DM respond to the potential, and the radiation pressure is absent so that in this regime, there is no gravitational distinction between matter and DM. In an RD epoch for fluctuations inside the horizon, there is only an approximate logarithmic growth of the DM perturbation which can be explored using the approximate logarithmic solution with dependence of the fluctuation on $K\tau$:

$$\delta(k,\tau) = A\Phi(0)\ln(BK\tau) \tag{8.1}$$

The approximate fit parameters to a full theoretical analysis are $A = 9.0$ and $B = 0.62$. This expression is plotted below in Fig. 8.3 for K in Mpc^{-1}, chosen by the user, as a function of $\alpha(t)/\alpha_{\text{eq}}$. The user can choose other values, but the formula has only a limited region of validity. A growth in δ means that matter is becoming more structured as time increases, as expected. The green vertical line indicates the scale factor at RD/MD equality. The slow growth occurs only approximately after the start of the

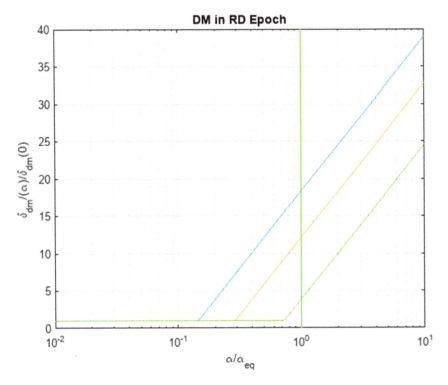

Figure 8.3: Dark matter logarithmic fluctuation as a function of $\alpha(t)$ from RD to MD epochs, $K \sim 10, 5, 2$ Mpc^{-1}. The blue line is for 2 Mpc^{-1}.

MD epoch. The user can choose the K value of the DM. The range of K values should cover only modes which have crossed/re-entered the horizon during the RD epoch.

```
% look at approximate CDM evolution for RD times
clear all
%close all
Ho = 1.0 ./4429 ;  % DH = 4429, Ho in Mpc^-1.
aeq = 1.47 .*10 .^-4; %  alpha - matter and photons only
%aeq = aeq .*1.68;  % 3 v generations;  % scale at RD/MD equality
omegm = 0.315 ; % critical density fraction of matter
Keq = Ho .*sqrt((2.0 .*omegm) ./aeq) % wave number at aeq,
K = alpha*H
```

```
Keq = 0.0148
```

```
zeq = 1 ./aeq  % z at RD/MD
```

```
zeq = 6.8027e+03
```

```
Kcdm = 2; %  K in Mpc^-1
a = logspace(-2,1,200); % dimensionless a/ao
% conformal time in RD
tauH = (2 .*0.95 .*a .*aeq) ./Ho ; % dimensionfull - Mpc , Hoto^0.95
B = 0.62;
del = 9.0 .*log(B .*Kcdm .*tauH);
```

8.3   DM Evolve: 2

The DM continues to evolve after the CMB decoupling, changing from a logarithmic growth during RD if inside the horizon to an approximately linear growing behavior during the MD epoch. Since it only interacts gravitationally, the perturbation will continue to increase with larger fluctuations as it continues to aggregate. A mode inside the horizon increases only as the logarithm during the RD epoch, while a mode outside grows as $\sim$ the square of the scale. Both modes in the MD epoch grow linearly with the scale factor (Table 8.1). For modes which have crossed back into the horizon well before t_{eq} the growth in the MD is linear as seen in the following. The DM perturbation for such modes satisfies the approximate differential equation (Meszaros equation):

$$[2y(y+1)]\frac{d^2}{dy^2}\delta_{dm} + (2+3y)\frac{d}{dy}\delta_{dm} - 3\delta_{dm}, \quad y = \alpha/\alpha_{eq} \qquad (8.2)$$

In particular, there is a linearly growing solution to this equation at large y of $\delta_{dm} = y + 2/3$. The solution to Eq. (8.2), done numerically here, in Fig. 8.4, should be smoothly matched to the logarithmic behavior seen in Eq. (8.1). However, that matching is not done here since the basic point is to illustrate the linear growth in the MD epoch, as shown in Fig. 8.2.

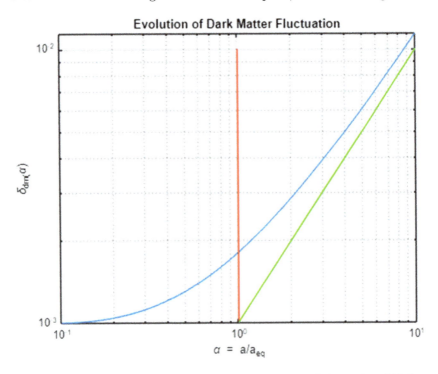

Figure 8.4: Growth of the DM fluctuation during the MD epoch for modes which have crossed the horizon early.

```
% use ode45 evolution  of DM pert in RD and MD
% DM pert far inside horizon
clear all
close all
% variable t is a/aeq
tspan  = logspace(-1,1,100); %
```

```
k = 1; % wave vector in Mpc^-1 - no explicit dependence - match to
log dependence
[t,y] = ode45(@mez,tspan,[0.0 , 1e-3]); % starting RD ,
evolve to MD [slope, position]
loglog(t,y(:,2));   % perturbation time development
title('Evolution of Dark Matter Fluctuation')
xlabel('\alpha = a/a_e_q')
ylabel('\delta_d_m(\alpha)')
grid
hold on
loglog([1 1], [1e-3 1e-2],'-r', [1 10], [1e-3 1e-2],'-g')
hold off
```

```
function dy = mez(t,y)
dy = zeros(2,1);
% evolution of dm perturbations
dy(1) = -y(1)*(2+3*t)/(2*t*(t+1)) + (3*y(2))/(2*t*(t+1));
dy(2) = y(1);
dy(2) = y(1); % 2 position, 1 velocity
end
```

The perturbation is flattened during the RD epoch and then rises rapidly, $\sim \alpha$, in the MD epoch. This type of behavior is expected for a Fourier mode re-entering the horizon during the RD epoch as mentioned previously. In the RD epoch, the fluctuation is growth suppressed due to the radiation pressure. In the MD epoch, the fluctuation grows more rapidly which results in an increasing "clumping" of the matter. The green line illustrates linear behavior.

The power spectrum derived from experimental data on galaxy surveys is shown below, in Fig. 8.5. Note that the convention is to normalize the power in units of Mpc^{-3}. This reflects the Fourier analysis of the two point correlation function normalized to a volume in k space as seen previously. There is both a linear behavior with physical wavenumber K and an inverse cubed behavior at larger K values. Small K values mean fluctuations not yet or newly within the present horizon during the MD epoch, while the earlier crossing modes have a characteristic inverse cubed behavior with K and have entered the horizon during the RD epoch. The quantum primordial fluctuations are taken to be Gaussian, with power $\sim K$. This behavior is seen at large scales, small K, because they are nearly outside the horizon at the CMB and retain a primordial behavior that $P(K) \sim K$. Smaller scales have re-entered at an RD or MD epoch and have been involved in causal

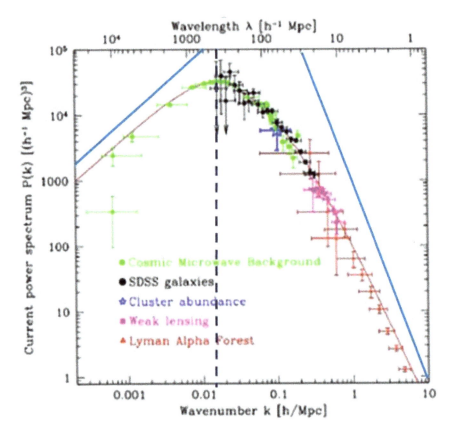

Figure 8.5: Power spectrum compilation from CMB and LSS data.

physics so that growth stops soon after re-entry at $K_{eq} \sim 0.015$ Mpc^{-1}. The falloff of the power spectrum is a crucial input dataset for the SMC parameter determination.

Large scale, small K, modes, primordial modes, grow as $K^{n_s} \sim K$ with this power normalization, while smaller scale modes are suppressed during the RD epoch by a factor $\sim (K_{eq}/K)^4$ with respect to what they would have had they continued to grow, as seen in Fig. 8.2. A maximum would then occur near the RD/MD equality and the power would turn over at larger K with a net $1/K^3$ behavior, $K(K_{eq}/K)^4$. It should be noted that the linear theory used here will break down at scales of $K \sim a$ few Mpc^{-1}, and a much fuller nonlinear treatment will be needed. However, the gross features of the power spectrum are understood. The blue lines are primordial K and RD suppressed K^{-3} scaling. The dashed black line is K_{eq}. The left boundary of the plot is $\sim K_o = 0.00025$ Mpc^{-1}.

The success of the CMB analysis confirms that the quantum fluctuations of inflation are small, $\delta T/T \sim 10^{-4}$. This fact implies that linear perturbation theory is applicable. As time advances, the fluctuations are, however, unstable gravitationally. Structure growth is expected in matter because while the CMB freestreams, the matter interacts. A Newtonian analysis results in a Poisson equation with radiation, matter, and dark energy sources. Since the Universe has for most of the time to the present been in a MD epoch after the short RD epoch, the LD effects can be ignored in the first approximation. The charged "baryons" can also be ignored with respect to the effects of dark matter, $\Omega_m \sim 0.315$, $\Omega_b \sim 0.05$. The gravitational potential is driven by the matter fluctuation as the Poisson equation source.

For modes which are suppressed in the RD, epoch growth occurs only after t_{eq}. Structure growth is frozen in the RD due to radiation pressure. The equation is approximately $\frac{d^2}{dt^2}\delta + 2H\frac{d}{dt}\delta - (3/2)H^2\Omega_m\delta$. In the RD epoch, the constant term is small and the equation simplifies to $\frac{d^2}{dt^2}\delta + 2H\frac{d}{dt}\delta \sim 0$ with solutions $\delta(t) = c_1 + c_2\ln(t)$ which is approximately frozen with only very slow growth. In the MD epoch with $a \sim t^{1/2}$, $H \sim 2/(3t)$, and $\delta(t) \sim c_1 t^{2/3} + c_2/t$, there is a growing mode $\delta(t) \sim a(t)t^{2/3}$. For completeness, in the ΛD epoch, $\delta(t) = c_1 + c_2 e^{-2H_\Lambda t}$, which approaches a constant, as fluctuation growth slows in a ΛD epoch.

Assuming a flat, matter-dominated Universe where photons and "baryons" are ignored as minority sources, a neutral non-relativistic (cold, neutrinos also ignored here) fluid called the CDM model can be explored. As is seen in Jean's mass discussion to follow, small structures form first, and then larger ones aggregate, with fluctuations scaling approximately as $\delta \sim a(t)/(M^{2/3})$. In the RD epoch, δ grows for modes outside the horizon as a^2, but modes inside the horizon are frozen by the photon pressure. In the MD epoch, radiation is unimportant and all modes grow with the expansion as $\delta \sim a$. The "primordial" modes have a power $P(K) \sim K$ assuming quantum scale-free Gaussian fluctuations.

Fits made with this simplified model (BBKS) interpolate between the primordial $P \sim K$ and the small-scale radiation suppressed behavior $P \sim 1/K^3$. In the script, the user picks the value of Ω_m and can view how the matter affects the power at small scales, with results shown in Fig. 8.6.

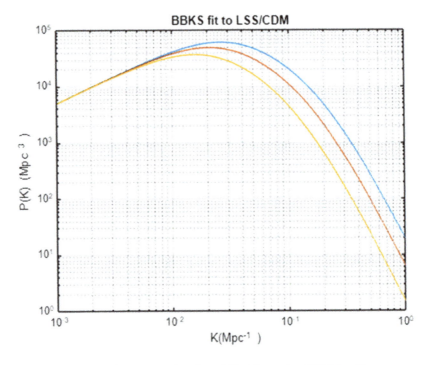

Figure 8.6: Power as a function of K with varying CDM densities.

```
c1 = 2.34; c2 = 3.59; c3 = 16.1; c4 = 5.46; c5 = 6.77;
% pick  omega m
OMm = 0.3;
% populate K in 1/Mpc
K = linspace(0.001,1,5000);
q = K./OMm;
a1 = log(1+c1.*q)./(c1.*q);
a2 = 1+c2.*q + (c3.*q).^2 + (c4.*q).^3+(c5.*q).^4; a2 = (a2).^-0.5;
T = a1.*a2;
PK = K.*T.*T; % norm using CMB
PK = PK .*5e6;
loglog(K,PK)
%
grid
title('BBKS fit to LSS/CDM')
xlabel(' K(Mpc^-1 )')
ylabel('P(K) (Mpc ^3 )')
hold on
```

The hypothesis of dark matter was part of the SMC because, among other things, the galaxy surveys had much more "clumping" earlier than purely matter could explain. Dark matter was needed to accommodate the rapid formation of structure in the Universe. As seen below, Fig. 8.7 (left) DM extends the power to larger values of K than does hot DM or a Matter + CDM mixture. There is no DE assumed. On the right, the present SMC mix with ~30% cold dark matter with only ~5% ordinary matter is the result of such fits. One might ask, where did the CMB oscillations go? Because the matter does not free stream, the CMB oscillations are suppressed and are, at present, unobservable in the matter power spectrum due to measurement systematic errors.

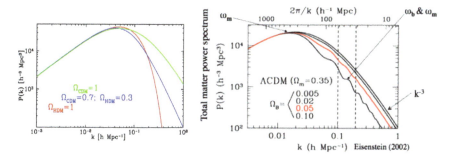

Figure 8.7: (left) Predictions in the SMC for a cold DM, hot DM mix. (right) The SMC but with different "baryon" densities.

In the evolution of structure, neutrinos can act as hot (relativistic) dark matter (HDM) depending on their exact masses, thus modifying the (non-relativistic) CDM model since they decouple when relativistic, while WIMPs are CDM since they decouple while non-relativistic. Neutrinos contribute to $\rho_{r_{el}}$ with the photons, and modify the evolution of the RD epoch. The neutrinos free steam which reduces the power at small scales because HDM tends to wipe out the DM density fluctuations. For small K, large scales, the power spectrum remains primordial. For larger K, smaller scales, the power grew only when the wavelength Λ was less than D_H so that fluctuations are frozen until t_{dec} and power is suppressed.

HDM also tends to suppress small scales by making the power turnover occur at smaller K due to the change in t_{eq}. The absence of such effects allows a cosmological limit to be placed on the neutrino masses. This limit is very competitive with laboratory limits and future surveys will improve on the limit. The synergy of the research of high energy physics (HEP) and Cosmology illustrates how the two efforts are complementary and reinforce one another. To go further into LSS requires that the mix of matter, DM, and photons be considered.

8.4   Jeans Collapse MD

If a local matter fluctuation exceeds a limit, that may cause a local region to collapse under gravity leading ultimately to star formation. Note that the timescales here are $\sim$1000 Myr, and larger mass collapses are quite slow. The spatial scales are on the order of 1000 Mpc. The matter is in a state of unstable gravitational equilibrium where small changes grow. Consider the matter at temperature T and with a perturbation of physical, not comoving, size λ. There are two competing effects: the thermal energy, U_T, and the gravitational energy, U_G. This is the temperature of the matter, not the radiation field since the MD epoch is being considered. For a small perturbation, the gravitational energy change is small, and there are sound waves due to the thermal energy stabilizing the perturbation. For larger λ, $U_T \sim kT$, $M \sim \rho\lambda^3$, $U_G \sim GM^2/\lambda \sim GM\rho\lambda^2$, and the ratio increases with the wavelength of the perturbation, $U_G/U_T \sim \lambda^2$. So gravity will dominate at larger scales. The Jeans mass is defined to occur when the effects are equal. Below the Jeans mass, pressure dominates, and the perturbation will re-expand when compressed by the sound wave. However, above the Jeans mass, gravity wins, and the perturbation collapses. Note the use of G here rather than M_p^2 because this is purely classical physics. Also, the physical wavelength is here called λ, also by convention rather than using Λ which is the convention adopted elsewhere in the text:

$$U_T = MkT/(\gamma - 1)\mu m_H, \quad U_G = -(3/5)GM^2/\lambda \qquad (8.3)$$

The parameter γ is the ratio of the specific heats of the gas, 5/3 for a mono-atomic gas, and μ is the mean molecular weight of the gas. The value is 1 for neutral hydrogen and 0.5 for fully ionized hydrogen. A spherical uniform distribution of the matter is assumed. This is purely a classical

analysis, simply in order to get a feeling for the orders of magnitude. The
critical perturbation length occurs when $2U_T = |U_G|$. The critical Jeans
mass and wavelength are

$$\lambda_J \sqrt{5KT/2\pi(\gamma - 1)\mu m_H G\rho}, \quad M_J = (4\pi/3)\rho\lambda_J^3 \qquad (8.4)$$

The Jeans wavelength is proportional to the sound velocity, c_s, which
scales as $\sqrt{T}$, whereas the collapse time, t_c, scales as $\sim 1/\sqrt{\rho}$. Since the
pressure must respond to stabilize over a distance $\sim \lambda_J$, it is easy to understand that $\lambda_J = 2\pi c_s t_c$.

$$t_c = \sqrt{3\pi/(32G\rho)}, \quad c_s = \sqrt{\gamma kT/m} \qquad (8.5)$$

The Jeans mass scales as the 3/2 power of T and inversely as the square
root of the density. As a cloud starts to collapse, the density increases,
and T will rise. If a smaller cluster existed inside the cloud, masses would
also start to collapse and the cloud may then fragment. Indeed, if $T^{3/2}$
increases slower than $\rho^{1/2}$, fragmentation will occur, which will if $\gamma < 4/3$.
If the radiation cooling can maintain the gas as isothermal, $\gamma = 1$, and fragmentation is expected. This means cooling is important for the formation
of stars. Typically, the Jeans mass is many solar masses, and a mechanism
is needed to understand why stars are so much lighter. Earlier, in Section 8.2, the understanding of the lifetimes of stars seemed to be in hand,
except that how they were initially formed was left open. The standard
explanation of stellar sizes is the fragmentation of the large collapsing gas
clusters.

The collapse time scales as the inverse 3/2 power of the fluctuation. In
this classical treatment, the formation of compact objects like stars takes
hundreds of millions of years. However, the newly commissioned Webb telescope has already found, in 2022, a complete galaxy which is only 390 Myr
old. The understanding of galaxy and star formation is not yet complete.
Indeed, most galaxies have a supermassive black hole at their heart and
how that formed so quickly is also an outstanding question. Looking at
stars orbiting these objects, one can estimate their masses. They range
from 10^5 to 10^{10} solar masses. Without cooling and fragmentation, they
might be formed directly in a collapse of a massive fluctuation. However,
the stars themselves are much less massive and are thought to have been
fragmented to smaller sizes with cooling. So how did the supermassive black

holes escape the fragmentation by which the collapsing objects arrived at solar masses? The classical explanation (Chapter 2) is that the stars form, burn out, and collapse into $\sim$ solar mass black holes which then somehow coalesce into supermassive size. However, the timescales for those processes are much longer than the Webb observation of supermassive black holes in the early Universe. Clearly, LSS is a very active area of inquiry, with new data to be confronted by model builders.

Given the issues, the following numerical exercise is as simple as possible. The CMB density fluctuations are small, $dT/T \sim 10^{-5}$. Decoupling occurs at $z_{\text{dec}} \sim 1100$ and the photons free stream uncoupled from the hydrogen and helium. Over-dense regions of matter can collapse directly. The collapse is opposed by gas pressure at a temperature T. As the fluid expands and cools, collapses are triggered. Large, cool, and over-dense regions which exceed the Jeans length will collapse. Smaller regions will oscillate with pressure waves. The Jeans length is the smallest size that collapses, now approximated as $\lambda_J = \sim\sqrt{kT/G\rho m} \sim c_s/\sqrt{G\rho}, c_s \sim \sqrt{kT/m}$. The timescale for collapse is estimated by ignoring pressure and using the free-fall time, $t_c \sim 1/\sqrt{G\rho}$, and all sizes have the same collapse time in this simplified analysis. The Jeans mass scales as $\sqrt{T^3/\rho}$.

```
close all
clear all
% constants
To = 2.7; rhoo = 1e-28 ; % present temp and density, scale as 1/a
and 1/a^3
Td = 3000 ; adec = 0.00091; rhod = 1.4e-19; % decoupling T, scale
and density.
% rhod = 2Mo/(pc)^3
pc = 3.2e16; k = 1.4e-23; G = 6.7e-11; mp = 1.7e-27; % 1 pc (m), kb
(J/K),
Mo = 2e30; Myr = 3.15e13 ; % solar mass, Myr sec
LJ = sqrt(k* Td/(G*rhod*mp));
LJpc = LJ/pc  % Jeans length at dec in pc
```

```
LJpc = 50.7164
```

```
MJ = rhod .*LJ^3;
MJsol = MJ/Mo % Jeans mass in solar masses
```

```
MJsol = 2.9922e+05
```

```
tc = 1/sqrt(G*rhod);
tc = tc ./Myr   % collapse time in Myr
```

```
tc = 10.3654
```

The estimates for the first collapsed objects are of the size of globular clusters, expected to appear after ~10 Myr. Stars are too small to collapse directly, and a "dark age", Fig. 6.4, is expected before stars begin to collapse and ignite. Since in the MD epoch T scales as $1/\alpha$ and $\rho \sim 1/\alpha^3$, $\alpha \sim t^{2/3}$, the Jeans mass is ~ constant, the Jeans length scales as $\lambda_J \sim \sqrt{T/\rho} \sim \alpha \sim t^{2/3}$, and the collapse time scales as $t_c \sim 1/\sqrt{\rho} \sim \alpha^{3/2} \sim t$ so that as cooling increases, the collapse takes longer. For scales $\alpha \sim 0.01$, the Jeans distance scales are ~0.5 Gpc with collapse times ~0.5 Gyr. However, the understanding of early LSS formation is presently evolving rapidly, so the classical Jeans analysis should be taken to give order of magnitude estimates at best. The Webb telescope will explore to a $z \sim 20$, or a scaled radius $a(t)/a_o \sim 0.048$. This is still well above the value for CMB decoupling however. Collapse times and masses as shown as a function of in Fig. 8.8.

```
al = linspace(adec,0.1); % scale Jeans length  and collapse time to
the present
LJpcs = LJpc .*(al ./adec);
tcsc = tc .*(al ./adec) .^1.5;
subplot(1,2,1)
loglog(al, LJpcs,'-b')
hold on
grid
title('Jeans Length in Mpc')
xlabel('\alpha')
ylabel('L_J (Mpc)')
loglog(0.0018,1000, 'b*', 0.05,1000, 'r*' )
hold off
subplot(1,2,2),
loglog(al, tcsc, '-b')
hold on
grid
title('Collapse Time in Myr')
xlabel('\alpha')
ylabel(' t_c(Myr)')
loglog(0.0018,1000, 'b*', 0.05,1000, 'r*')
hold off
```

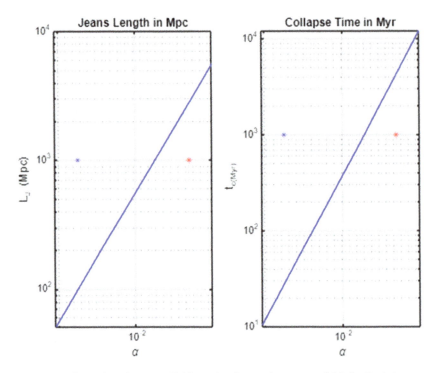

Figure 8.8: Jeans length vs. α (left), and collapse time vs. α (right). Both increase with α or t. The stars indicate $2\alpha_{\text{dec}}$ (blue) and the Webb range $\sim \alpha_W = 0.05$ (red).

These analyses are classical in a static space, without Hubble expansion, and are vastly simplified. A more correct supercomputer simulation is shown in Fig. 8.9. Although at $z \sim 30$ the matter density is quite uniform on the distance scale chosen, nonlinear effects set in rapidly. It is interesting to note that the closed over-density of a collapsing region shown in Section 3.7 has been used to make first-order nonlinear corrections to modify the fits shown in Section 8.3. Clearly, the nonlinear nature of the problem of galaxy structure requires much more advanced tools than are deployed in this text. Indeed, it is a large current research activity in itself. In addition, there are unanswered questions in regard to the early existence of supermassive black holes which need to be reconciled with the models. Early supermassive black holes are exciting new data with which to confront the existing models.

Recently, astronomers have begun to observe a different type of galaxy: a "dark galaxy", with an example shown in Fig. 8.10. These objects appear

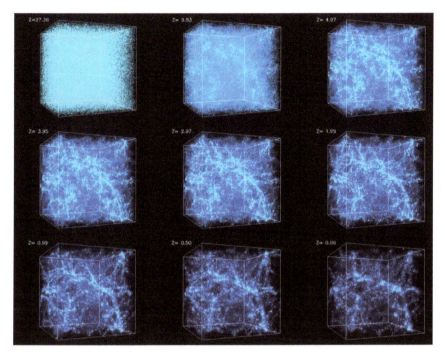

Figure 8.9: Supercomputer simulation of LSS evolution with snapshots at different z values.

to be composed of dark matter and ordinary matter in the form of hydrogen. They are observed using radio telescopes and viewing the hydrogen emission spectrum. The matter seems not to have collapsed into stars but remains a gas of hydrogen and dark matter. Clearly, such objects are not part of the standard scenario for galaxy formation. The results of further searches and the response of model builders for LSS will be of great interest.

There are many new surveys of galaxies in preparation. Data are presented in many ways which the reader should be conversant with. The preferred unit of length for discussions of structure is the parsec which is about 3.26 light years (lyr), 2.05×10^5 AU or 3.08×10^{18} m. This is a parochial unit defined by the size of Earth's orbit which yields a parallax second on a baseline of 2 AU at a stellar distance of 1 pc. The nearest star to Earth is at 1.2 pc. Following Newton, one can assume all stars are like the Sun and note that the nearer stars are about 10^{11} less bright. Knowing the distance to the Sun and assuming an inverse square law, the distance

Figure 8.10: A "dark galaxy" with few stars but with a fairly typical mass and size.

to the nearest star is then estimated to be ~1.5 pc. Newton did the first survey!

Assuming a present horizon of three times D_H, and a mean distance between galaxies of 1.5 Mpc, the total number of possibly visible galaxies is ~10^{12}. For a galaxy of radius r_g and mass M_g equating gravitational attraction and centrifugal force, the velocity of galactic rotation is $v_g/R_g = M_g/(M_p r_g)^2$. The Milky Way rotates at a velocity ~200 km/sec (Doppler shifts) and is about 0.02 Mpc in diameter which means the mass is about 10^{11} solar masses.

8.5   Dark Matter: Baryon Acoustic Oscillations

There now exists the very successful standard model of cosmology (SMC). It consists of dark energy, dark matter, and an inflationary scenario. The age of the Universe is 13.82 Gyr in this model. The dark matter was added to account for galactic rotation curves and the rapid clustering of matter in

LSS. Dark matter is also independently inferred from lensing observed in galactic surveys. The density fluctuations are initially amplified by gravitational instabilities. The Hubble expansion was measured, using supernovae as standard candles, to be slowing. These data required the addition of a repulsive cosmological source: the dark energy. As a final element, the smoothness of the CMB required an early inflationary period to bring the CMB quickly into causal contact in order to make the CMB spatially uniform.

There are no accepted candidates for these crucial SMC elements, DM and DE, and the field responsible for inflation. The DM was looked for as WIMPs, but the present limits have nearly hit a floor due to irredeemable neutrino backgrounds. For this reason, DM searches have been altered to encompass lower mass and more exotic candidate, axions, sterile neutrino, etc. A next generation WIMP search is proposed now to reach the neutrino floor and perhaps beyond it. The DM has so far been assumed to be stable and to have only weak interactions. However, it could possibly slowly decay into detectable SMPP particles and searches are also ongoing for those types of signatures. For dark energy and the cause of inflation, no SMPP candidates exists, except perhaps the Higgs which must be coupled to gravity in a non-standard fashion to fit the inflationary scenario. So, the SMC asserts the existence of objects currently unknown. Indeed, the SMC is, as it should be, a working model, the simplest paradigm that fits all the present data, within errors.

However, the data continue to improve. One crucial check of the SMC is the measurement of baryon acoustic oscillations (BAO) that have recently been obtained. The CMB has multiple acoustic peaks caused by the sound oscillations present in the photon–baryon fluid at decoupling. The photons have free streamed, approximately, since then. However, these same oscillations must exist in the tightly coupled baryons. However, the matter has certainly not free streamed since then (Jeans collapse, for example). A cartoon of the CMB/BAO correspondence appears below, Fig. 8.11, which indicates a BAO background of highly structured galaxies. The BAO "standard ruler" is closely related to the CMB acoustic peaks.

Data from galaxy surveys of millions of galaxies have become available, and newer instruments are being constructed which will come online soon. The Sloan Digital Sky Survey (SDSS) initially identified the BAO fundamental acoustic peak. The subsequent Dark Energy Survey Instrument (DESI) has greatly improved these data, thus sharpening the comparison to the CMB data. Exploration of these structures in a survey uses the

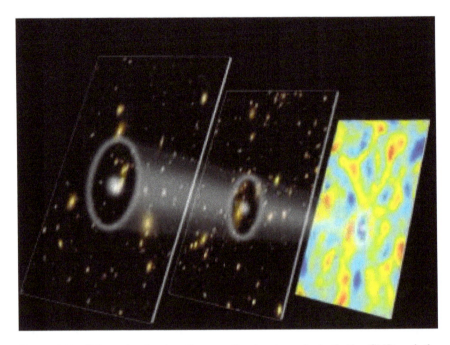

Figure 8.11: Schematic showing the acoustic structures in both the CMB and the galaxies.

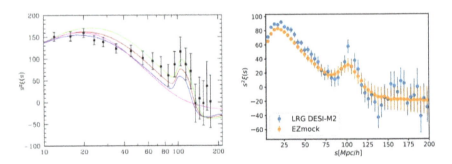

Figure 8.12: Data from the SDSS (left) and the DESI (right) showing the fundamental acoustic peak — the BAO.

correlation functions which are plotted below in Fig. 8.12. They mirror and quantify the rings displayed in Figure 8.10.

What is measured in the BAO surveys is the correlation enhancement effect and z to scale to the present. This distance is effectively a "standard ruler". The length of the ruler, s, is the distance a sound wave can travel from the HBB to the photon–baryon last scattering which is very approximately, ignoring baryons, $c_s \sim c/\sqrt{3}$ and $s \sim c_s t_{dec}$ is the "sound horizon". That crude estimate is $s \sim 78$ Mpc scaling to the present using z_{dec} which depends on H_o. The "last scattering" time should be used instead of the decoupling time. Indeed, the BAO analysis can measure $H(z)$ with sufficient z range contained in the survey. The accepted value of s is 147 Mpc, $\sim 100/0.68$, and the data are used to determine $H(z)$ for the survey z range. This method to find H is independent of and distinct from using a series of "standard candles" — Cepheid variables to type 1 supernovae to find the Hubble constant.

The "standard candle" method uses the luminosity distance, while the standard ruler uses the angular size, D_A, defined in Section 3.3. In fact, there is a small tension between the two methods of measurement: CMB and BAO. The CMB data fit yields a measurement of H_o as a parameter. A "direct" measurement of H_o using supernova data as a distance scale using other methods is in disagreement with the best fit values by, at present, $\sim 9.8\%$. A comparison is shown in Fig. 8.13. The contours shown in Fig. 8.13 are for 1 and 2 standard deviation likelihoods (68% and 95% confidence contours). This may be a serious discrepancy, or it may be a statistical fluctuation. Another possibility is that the complex methods, vastly oversimplified in this text, needed to extract the SMC parameters from the two datasets may not yet be complete. Time and improved data and analysis techniques will tell.

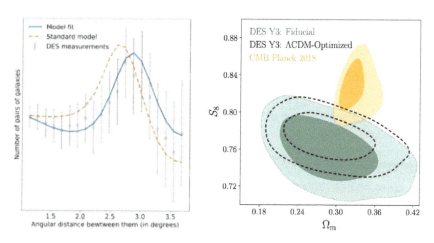

Figure 8.13: Angular distance of BAO peak compared to SMC (Planck) (left). Comparison of allowed contours for SMC parameters using CMB and DES data (right).

8.6   Issues BAO/CMB, New Data

Inflation is basically unique in that it predicts CMB perturbations in TT, EE, and BB mode pairs of the proper magnitude in distinction to almost any other theory. The CMB acoustic oscillations are correctly understood and help define cosmological parameters, providing cross-checks to supernovae measurements of DE and nucleosynthesis limits on baryon density and the number of neutrinos. The present cosmology SM is simple and self-consistent and explains a vast amount of ever improving precision data. The BAO confirms the baryon oscillations and checks the inflationary model in yet another fashion. There are still other predictions to confirm. For example, the BB modes and gravitational spectral index remain to be observed. New more precise data will improve on present limits or, perhaps, uncover new phenomena beyond the present SMC. Indeed, new Webb telescope data may raise issues in LSS models which in turn might raise questions for the SMC.

More data are clearly needed, in connection with LSS, especially if the modest tension between the CMB and BAO analyses persists and sharpens. For example, it has been found that most galaxies have a supermassive black hole at the core. How did they form and when? More detailed modeling may shed a different light on the observed galactic structures. Another issue is the formation of early galaxies. To confront the new Webb data, one must probably make a real model of DM. Indeed, the new data will be a testing ground for explicit comparisons with specific models of DM: heavy, light, cold, hot, etc. The model in Section 8.2, that black holes were not much more massive that the sun and that they formed only after stars burned out, on a scale of billions of years, is obviously not the whole story.

One recent attempt to obtain new and different data has been to look at the modulation of the timing of pulsars in order to identify the passage of the gravitational waves of primordial black holes. Inspirals like the binaries are already discussed but now on a much vaster scale, have a cartoon shown in Fig. 8.14, with first reported data displayed in Fig. 8.15. Or the gravitational waves generated by inflation can be searched for. The baseline for these searches must be immense given the present sizes of the radiating objects. The dimensionless amplitudes h are expected to be of order 10^{-15} with frequencies of order $1/\mathrm{yr}$ or nHz for masses in the range of supermassive black holes, 10^5–10^{10} solar masses. Basically, the GR wave modulates the distance of the free fall pulsars on the slow timescales which are dictated by the extreme masses of the radiating sources. These data will open a window on the very early formation of galactic structures.

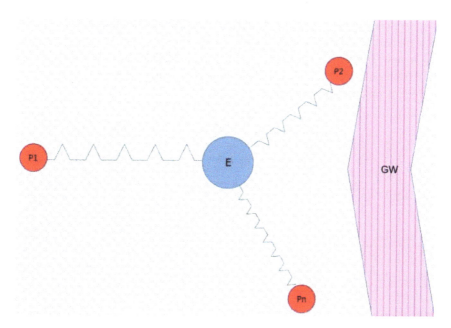

Figure 8.14: Cartoon of the use of pulsar timing modulation to detect very long wavelength gravitational waves.

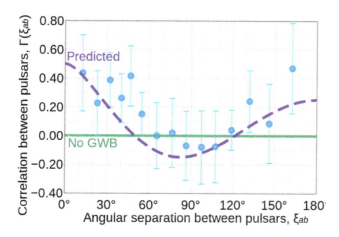

Figure 8.15: First reported signal evidence for GR waves using modulated pulsar timing.

Proposed detectors of GR radiation are shown below in Fig. 8.16. LISA is a space-based interferometer with a very long baseline which would open up a new range of frequencies, $\sim 10^6$ times smaller frequencies than the Earth-based detectors. The pulsar timing in turn opens up another ~ 4 orders of magnitude in frequency.

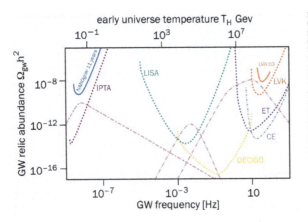

Broadband Sensitivity of current (solid) and future (dashed) gravitational-wave (GW) observatories to stochastic GW backgrounds (expressed in terms of the energy density fraction in the universe today). On the upper x-axis, the temperature in the early universe is given, which is obtained when the peak frequency of a GW signal is equal to the inverse of the Hubble expansion rate when GWs are emitted. Some example possible GW spectra from the early universe are also shown (pink, dashed). Source: F Rompineve/arXiv:2101.12130/arXiv:200

Figure 8.16: Projected sensitivity contours for GR wave detectors at low frequencies, well below the LIGO sensitivity range which is $\sim$kHz, not the $\sim\mu$Hz range and below.

Another long-standing issue is simply the existence of matter in the Universe. Why not just radiation and all particles annihilate with their antiparticle? The conditions for matter to exist are that there be a period of time out of thermal equilibrium, that reactions that violate CP conservation exist, and that there are reactions that violate baryon number. The first two conditions are known to be met in the history of the Universe, but the amount of CP violation currently measured in the SMPP is too small. However, CP violation in the neutrino sector of the SMPP has still not been well measured, although several new experiments will come online soon. An additional constraint on the models beyond the SMPP is that protons not only exist but are also immensely stable. Table 8.2 below shows the current limits and the predictions of several extensions of the SMPP. Most notably, many SUSY models are ruled out. These models are largely variations of supersymmetric models, but WIMP searches and LHC direct searches have also ruled out some of these models. The scale of the proton lifetime limit is currently about 10^{35} yr, or almost 10^4 times longer than the age of the Universe. How can matter exist and yet be so stable? Perhaps the relevant reactions have an extremely steep energy dependence or a threshold....

Table 8.2: Limits on proton lifetime for several postulated theoretical models, GUTs or SUSY.

Theory class	Proton lifetime (years)[18]	Ruled out experimentally?
Minimal SU(5) (Georgi–Glashow)	10^{30}–10^{31}	Yes
Minimal SUSY SU(5)	10^{28}–10^{32}	Yes
SUGRA SU(5)	10^{32}–10^{34}	Yes
SUSY SO(10)	10^{32}–10^{35}	Partially
SUSY SU(5) (MSSM)	~10^{34}	Partially
SUSY SU(5) – 5 dimensions	10^{34}–10^{35}	Partially
SUSY SO(10) MSSM G(224)	$2 \cdot 10^{34}$	No
Minimal (Basic) SO(10) – Non-SUSY	$< {\sim} 10^{35}$ (maximum range)	No
Flipped SU(5) (MSSM)	10^{35}–10^{36}	No

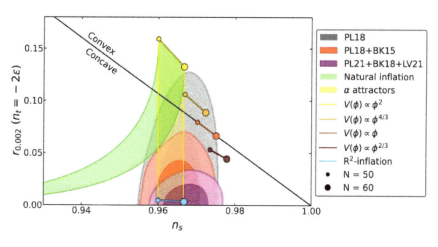

Figure 8.17: Compilation in 2024 using Planck, BICEP/Keck, and LIGO/VIRGO data.

As the data improves, with a present compilation shown in Fig. 8.17, more incisive constraints are possible. For example, the evolution of the spectral index with k has contributions from the third derivative of the potential. Using all the present (2023) data, a recent compilation and fit found that, at the 95% CL, $r < 0.028$ and n_s between ~ 0.958 and 0.977. Clearly, power-law potentials are now quite disfavored.

Chapter 9

Higgs and the SMC

"We shall not cease from exploration, and the end of all our exploring will be to arrive where we started, ... A condition of complete simplicity .... And all shall be well."
— Little Gidding, T. S. Eliot

"The aim of particle physics is to understand what everything's made of, and how everything sticks together. By everything I mean me and you, the Earth, the Sun, the 100 billion suns in our galaxy and the 100 billion galaxies in the observable universe. Absolutely everything."
— Brian Cox

"The standard model of particle physics describes forces and particles very well, but when you throw gravity into the equation, it all falls apart. You have to fudge the figures to make it work."
— Lisa Randall

The SMC and the SMPP are the two present major theoretical constructs in fundamental physics. Both have many open questions and issues, but that is what a vital scientific enterprise looks like. For the SMPP, there is an issue of internal self-consistency which arose with new and more precise data. There are also issues with the newly discovered Higgs boson itself. In addition, there are many, many parameters which are not predicted but are known only by measurements. Why does the Universe have that particular set of constants?

9.1 SMPP: Self-Consistency

Since the Higgs gives mass to all the SMPP particles, it must be consistent with the specific measured W boson and top quark mass. In addition, the model should be consistent with an end to slow roll inflation, followed by an oscillating field that produces the SMPP particles and reheats to form an HBB starting with an RD epoch and, finally, yielding the observed CMB fluctuations. How the reheating happens, in detail, is an open question for model builders to explore.

The first question to ask is perhaps whether or not the SMPP by itself is internally self-consistent. The Higgs gives mass to all SMPP particles but couples to mass, which means the effects are largest for the heavy W and Z bosons and the top quark. Recently, new measurements of the W mass at the Fermilab Tevatron and the LHC have been reported as well as more refined Higgs mass values as the data logged at the LHC continue to increase and the analyses improve. The LHC also produces copious quantities of top quark pairs, and new results are also becoming available. In fact, they are giving some tension from the combination of the masses of the Higgs, top, and W, as seen in Fig. 9.1. As per normal, likelihood contours of 1 and 2 standard deviations are plotted which illustrate the modest tension in the SMPP.

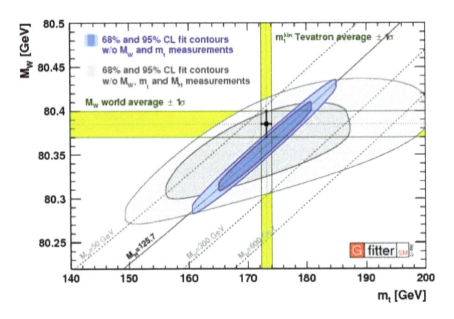

Figure 9.1: Plot of the SMPP consistency for the W, top quark, and Higgs mass.

The SMPP has three symmetry groups: one each for the strong force, $SU(3)$, the weak force, $SU(2)$, and the electromagnetic force, $U(1)$. The force carriers — gluons, W and Z, and photons — each have a coupling constant, α whose value is determined experimentally. The vacuum expectation value of the Higgs field, $\langle\phi\rangle$, gives mass to the Higgs, W, and Z. Numerically, there is a parameter θ_W, the Weinberg angle, whose measured value is $\sim 29^o$ and which relates the W and Z mass. The Higgs, W, and Z masses are all related in the SMPP in a unique fashion. The Higgs and Z masses are quite precisely determined while the W and top quark masses are somewhat less well measured due to the presence of undetected neutrinos in the W decay processes and the large decay width of the top quark. The Higgs mass was previously shown in Eq. (2.24). The constant g_w is the weak coupling constant, analogous to e:

$$M_H = \sqrt{\lambda}\langle\phi\rangle, \quad M_W = g_W\langle\phi\rangle/\sqrt{2}, \quad M_Z = M_W/\cos(\theta_W) \tag{9.1}$$

The angle, θ_W, mixes the weak and electromagnetic forces. It is also an experimentally determined parameter of the SMPP. The weak Fermi coupling, G_F, is defined by the measured value of the weak decay rate of muons. The parameter λ in the Higgs potential can only be determined by experimentally measuring the Higgs mass:

$$\Gamma_\mu = G_F^2 m_\mu^5/(192\pi^3), \quad \langle\phi\rangle = 1\Big/\left(\sqrt{G_F\sqrt{2}}\right) \tag{9.2}$$

There are higher-order corrections to the W mass from the top and Higgs that alter the prediction for the W mass. These corrections are due to quantum field theoretic effects and are called the RGE or renormalization group equations. At the lowest order, due to virtual particles, the W mass is shifted with respect to the Z mass. The loop charges may screen or anti-screen the central charge. The mix of particles in the cloud at a given distance from the charge changes with energy because the virtual particles have a range of order in their Compton wavelength which is defined by their mass. Therefore, the effective screened charge depends on the energy at which it is probed. To the lowest perturbation order, the virtual top and Higgs shift the mass of the W with respect to the Z:

$$M_W = M_Z \cos(\theta_W)\sqrt{1 + \Delta_t + \Delta_H}$$

$$\Delta_t \sim 3\alpha_W(m_t/m_w)^2/16\pi, \quad \Delta_H \sim -(1 + \alpha_W \tan^2(\theta_W))\ln(m_H/m_w)/(24\pi) \tag{9.3}$$

New high statistics LHC data on the W mass, 09/2024, reduce the issues of self-consistency from what is shown above. For now, awaiting new data, it can be assumed that the SMPP is a self-consistent and complete quantum field theory description of fundamental particles at energies less than ~1 TeV. There are many *ad hoc* parameters in the SMPP and the existence of matter itself (baryogenesis?) is unexplained, but the SMPP, with the Higgs discovery, is now a complete description of the strong, weak, and electromagnetic forces.

9.2   Higgs Quartic Coupling

Another SMPP consistency check is to attempt to use the RGE to "run" the parameters of the SMPP to much higher mass, always assuming that no new physics intervenes at elevated mass scales. As the energy scale of the field increases, the quartic term in the Higgs potential will dominate. The quartic potential parameter, λ, runs with energy, in a way which depends critically on the top quark mass. This behavior argues for an effort to further experimentally constrain the top mass because of issues with the high energy stability of the theory. Issues with the results of running the couplings would be indicative of new physics intervening at high energies.

A very simplified representation of the RGE for the quartic Higgs potential at elevated temperature, T, depends on two parameters with virtual W, b_W, and virtual top particles, b_t and is:

$$\lambda_H(T) \sim \lambda(0) + \left(3/64\pi^2 \langle\phi\rangle^4\right) \left[3M_W^4 \ln\left(b_W T^2/M_W^2\right) - 4M_t^4 \ln\left(b_t T^2/M_t^2\right)\right] \tag{9.4}$$

The logarithmic factors result from the calculation of quantum loop integrals in the RGE procedure. As the energy scale, T, increases, the reduction in the quartic coupling constant is driven by the top mass term which increases. At some point, the quartic shape will change as shown in Fig. 9.2. For an energy of 10^{12} GeV, the Universe is predicted to have an unstable vacuum, which must indicate that new physics is coming into play during the factor $\sim 10^9$ extrapolation in the energy scale.

```
% look at finite temp field theory - 1 loop on Higgs lambda -
quartic
Mw = 80.4; Mt = 171; vev = 174;
bf = exp(1.14); b = exp(3.5);
dlam = 3 ./(128 .*pi .*pi .*vev .^4);
dlam = dlam .*(1.5 .*Mw .^4 .*log((b .*T .*T)/(Mw .*Mw)) - 2.0 .*Mt
.^4 .*log((bf .*T .*T) ./(Mt .*Mt)));
```

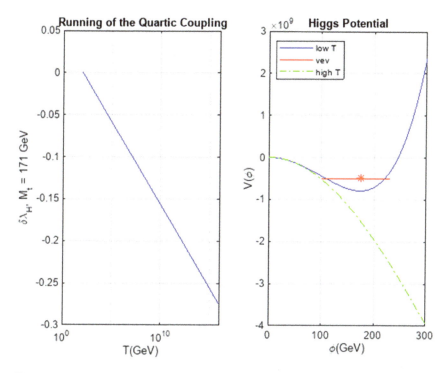

Figure 9.2: Running of the Higgs λ parameter with energy (left). Potential just as λ is driven down to 0, compared to vev at low T.

The steady decrease of the Higgs quartic coupling is notable in Fig. 9.2 (left) with a loss of a stable potential minimum (right). The current SMPP parameters, run up to 10^{12} GeV, lead to a nearly unstable vacuum. What that means is shown in a cartoon below. Clearly, this is not expected to be correct, so either the parameters of the SMPP need to be measured more accurately, or the SMPP needs adjustment, or — perhaps most likely — there is new physics between the $\sim$TeV scale currently probed and the billion times higher energy scale indicated. The present measurements indicate meta-stability, as in Fig. 9.3 (top) and the Higgs potential ceases to increase with the field, and, in fact, becomes negative, Fig. 9.3 (bottom). A cartoon of the effect is shown below.

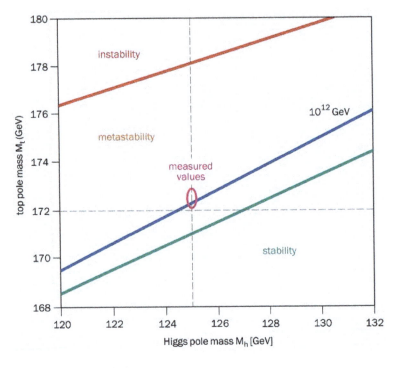

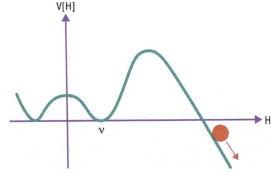

Figure 9.3: (top) Stability regions of the vacuum for a range of top and Higgs masses. (bottom) A graph of an unstable vacuum.

9.3 Higgs Inflation: Ricci Coupling

As a good physical theory, inflation makes testable predictions; the Universe is flat, density perturbations are almost scale invariant with a spectral index of almost one, there are acoustic peaks in the CMB and BAO, perturbations are Gaussian, and gravity waves exist the large scales Λ. Except for the last prediction, the others have been confirmed. The SMC has been well verified. However, there remains the issue that no known candidates exist for DM, DE, or the scalar field of inflation or its potential.

There is now, for the first time, a complete SMPP with a verified fundamental scalar field — the Higgs boson. Since inflation requires a scalar field, is it possible to use the Higgs? Such an identification would be economical and compact, so it is worth a try. However, this hypothesis assumes that there is no new physics from an energy of about $100\,\text{GeV}$–$10^{19}\,\text{GeV}$ which seems unlikely given past experience in physics. There would be no SUSY and thus no identified DM candidate, no GUT, and thus no explanation of the baryon asymmetry and the neutrino masses. New physics at or below the TeV mass scale, for example, SUSY, is needed in particular to have a particle physics candidate for DM. New physics at the GUT scale is indicated by the unification of the coupling constants and is often invoked to explain the baryon-antibaryon asymmetry of the Universe with a related prediction of proton decay. Nevertheless, the simplest proposal can be explored and any resulting contradictions can be evaluated.

Putting aside the vacuum stability issues while awaiting more precise measurements of the parameters of the SMPP, the possibility of a Higgs as the scalar field responsible for inflation can be considered as a minimalist approach. Extrapolating the Higgs potential to early times, the quartic part of the potential dominates. The Higgs will contribute to inflation but the Higgs mass of $125\,\text{GeV}$ is much less than the Planck mass, by ~ 16 orders of magnitude. However, a minimal set of assumptions is to try to connect the SMPP and the SMC with few, if any, added assumptions and explore the consequences. After all, the Higgs is the only fundamental object in the SMPP with vacuum quantum numbers, so it is worth an attempt.

A straightforward approach is to use the high mass part of the Higgs potential making the model a simple quartic potential model. The required strength of any potential to account for the CMB temperature anisotropy of about 10^{-5} is to have a large field, a large value of r, and a very small coupling constant $V(\phi)/\varepsilon = (3/8)P_H M_p^4 = (0.0054 M_p)^4$. Note that a simple quartic potential was found to be unstable, so the postulated Higgs potential must be more complex than the quartic term. The standard coupling for the Higgs will, therefore, not work. For example, $\lambda_H \sim 3.9 \times 10^{-14}$ is implied for a slow roll model. The power is an immediate problem in that with minimal gravitational coupling of the Higgs to energy, the coupling to match the CMB data is not the observed quartic Higgs coupling, not the value fixed by the Higgs mass. In order to cause inflation and act with a slow roll, the Higgs needs to be coupled in a non-standard way to the geometry as shown in the following, where R is the Ricci scalar, ϕ_H is the Higgs field, and ξ is a dimensionless coupling constant. The stress–energy tensor of normal GR, $T_{\mu\nu}$, does not work, and this alternative does:

$$\int d^4x \sqrt{-g} \left[- \left(M_p^2/2 \right) - \xi \phi_H^2 \right] R \qquad (9.5)$$

In the HBB evolution at temperatures above $\sim kT \sim M_Z c^2 \sim 10^{15} \, {}^\circ K$, $t \sim 10^{-10}$ s, all the known SMPP particles are essentially massless and the Higgs vev is zero (by assumption). When the temperature drops later in the evolution, the Higgs acquires a vev and the SMPP particles acquire their mass. However, this coupling does not result in a pure power law potential. Using the Ricci scalar coupling, the effective potential at SMPP energies remains as it was. However, at elevated energies, the effective potential becomes almost constant, which means that a slow roll solution has been constructed, in agreement with the CMB data:

$$V_H(\phi) \to V_H(\phi) / \left[1 + 8\pi\xi\phi^2/M_p^2 \right]^2 \to \lambda_H M_p^4/(8\pi\xi)^2 \qquad (9.6)$$

At high mass scales, the potential changes from quartic to a constant. The modified theory is a very slow roll. The cost is that the coupling must be large which may seem unnatural as does the non-standard coupling. The potential becomes approximately constant at the field value $\phi/M_p \sim 0.003$. Inflation will end at low field values with Ricci coupling $\psi/M_p \sim 0.05$ in this case. This means r is reduced easing the issues of quartic potential models being outside the limits set by the data. The mismatch of the Higgs λ_H parameter to the CMB power anisotropy is evaded because the potential at large field values is controlled by the non-minimal coupling parameter ξ. Using the required effective couplings of $\sim 10^{-13}$ and ~ 0.6 for the Higgs low mass quartic case at high and low mass scales and equating the two potentials, the coupling parameter is large. Approximate results for this model result in the predictions, $n_s \sim 0.966$ and $r \sim 0.0033$ for N_e of 60. The model is within the present experimental limits.

A comparison of the derived inflation parameters for two power-law potentials and the proposed Higgs potential is shown in Table 9.1. The CMB power normalization fixes the parameter ξ of the Higgs coupling.

Table 9.1: Comparison of Higgs inflation to two power-law potentials.

	Quadratic	Quartic	Higgs
V	$m^2\varphi^2/2$	$\lambda\varphi^4$	Non-minimal
$\varepsilon\pi x^2$, $x = (\varphi/M_p)$	$1/4$	1	
$\delta\pi x^2$	$1/4$	$3/2$	
$N_\Lambda/\pi x^2$	2	1	
$V/\pi\varepsilon$	$2\,(m/M_p)^2$	λx^2	
$(1 - n_s)\pi x^2$	1	3	
$N_\Lambda = 60 \to \varphi_i/M_p$	3.1	4.37	0.068
$1 - n_s$	0.031	0.050	0.034
$V/\varepsilon = \mathrm{CMB}\,\mathrm{d}T/T$	$(m/M_p) = 1.2 \times 10^{-6}$	$\lambda = 3.9 \times 10^{-14}$	$\xi \sim 47{,}000$
r	0.13	0.267	~ 0.003

The power-law potentials have simple analytic forms, explored previously. The Higgs results for the required $\sim$60 e-folds, the slow roll parameters, and the CMB normalization are shown in Table 9.1. For 60 e-folds, $n_s = 0.966$, $r = 0.003$. The Higgs "model" thus fits comfortably within the existing experimental constraints. In comparison, the quartic potential is disfavored due to the tensor parameter and the quadratic potential is in moderate tension, as seen in Fig. 8.16. In fact, the data prefer a slower roll than the power-law potentials. The user picks the initial Higgs field and the value for the non-minimal coupling constant ξ. The results for slow roll inflation are displayed. Figure 9.4 shows the potentials for a quartic model and the modified Higgs potential as a function of field. Figure 9.5 shows the slow roll Higgs parameters as a function of field. The number of e-folds is $\sim$39 for the initial settings which is a bit low, and the CMB power is a bit high. The user can choose other input parameters and see the effects.

The SMPP has determined the masses of the W, Z, top quark, and Higgs. They are all consistent with a VEV, which is now in effect. The Higgs minimum potential, VEV, is $\sim 10^{55}$ times (!) larger than the SMC value of the DE. Why the Higgs VEV is presently gravitationally inert is unclear. The ultimate reconciliation of the two scientific edifices describing fundamental physics is still a distant goal but one being most vigorously pursued. It seems clear, in any case, that the quantum vacuum is not well understood. Recalling that there is not a quantum theory of GR, this fact may not be surprising.

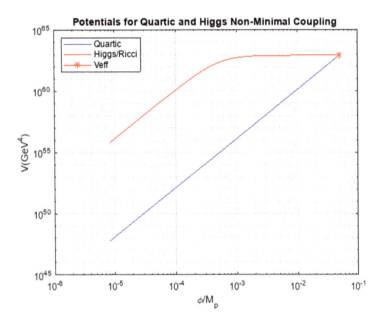

Figure 9.4: Higgs potential as a function of the field. A slow roll period evidentially exists.

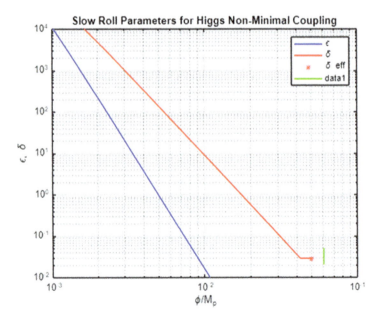

Figure 9.5: Slow roll parameters for the Higgs inflation model as a function of field.

```
% evaluate V for quartic inflation and for non-minimal Higgs
% also find inflation parameters for nmH numerically
Mp = 1.2 .*10 .^19;      % GeV - Planck mass
lamh = 0.6 ; % quartic coupling for Higgs - no RGE
lamq = 3.9 .*10 .^-14 ;  % coupling for quartic inflation
phii = 0.05; % Initial field/Mp
% Non-Minimal Coupling for Higgs
coupl = 160000 ; % Coupling
emin = 14; % log Minimum Field Energy Power (GeV)
emax = phii .*Mp; % phii was scaled to Mp
Vh = lamh .*(espan .^4) ./(1.0 + (8 .*pi .*coupl .*espan .*espan)
./(Mp .*Mp)) .^2;
Veff = (Mp .^4 .*lamh) ./((8 .*coupl .*pi) .^2)
```

Veff = 7.6941e+62

```
% Large Field Effective V
% find derivative and integrals numerically to evaluate nmH model
eps = (Mp .*Mp .*((dV ./Vh) .^2)) ./(16 .*pi);
eta = (Mp .*Mp .*(d2V ./Vh)) ./(8 .*pi);
ns = 1 - 6 .*eps + 2 .*eta;
r = 16 .*eps;
% strength wrt CMB dT/T
CMB = (((Vh ./(eps)) .^0.25) ./Mp) ./0.0054;
eps(100)
```

ans = 1.6426e-06

```
% Initial epsilon
eta(100)
```

ans = -0.0283

```
%'Initial eta
ns(100)
```

ans = 0.9434

```
% initial ns
r(100)
```

ans = 2.6281e-05

```
% initial r
CMB(100)
```

ans = 2.2702

```
% CMB Power Ratio
N
```

N = 39.2572

```
hold off
```

A non-zero limit on $n_s - 1$, Figure 8.17, with a very small ε value implies the current approximate limits on δ shown in green in Figure 9.5, consistent with the Higgs model values.

Appendix A

MATLAB Tools

MATLAB has a full suite of startup exercises and guides. There is the inline "help" query. In addition, the "search documentation" tab on the startup page allows the user to search for syntax and examples. Finally, there is a summary guide (pdf, partially enclosed here) of the many MATLAB operations that are available. A sample appears in Table A.1.

Using these tools, the user should easily follow the flow of the scripts that constitute the heart of the text. There are many "comment" lines which briefly explain what is being calculated or displayed in a given Live file.

Table A.1:

Operators and Special Characters	
$+, -, *, /$	Matrix math operations
$.*, ./$	Array multiplication and division (element-wise operations)
$\wedge, .\wedge$	Matrix and array power
$\backslash$	Left division or linear optimization
$', '$	Normal and complex conjugate transpose
$==, \sim=, <, >, <=, >=$	Relational operators
`&&`, $\|\|$, $\sim$, `xor`	Logical operations (AND, NOT, OR, XOR)
;	Suppress output display
...	Connect lines (with break)
`% Description`	Comment
`'Hello'`	Definition of a character vector
`''This is a string''`	Definition of a string
`str1 + str2`	Append strings

There is also a pdf file addressing connecting MATLAB and PYTHON script, which may be of use to the reader.

Appendix B

Power Law H

The R–W metric has power law solutions in cases where the energy content is dominated by matter, radiation, or a cosmological constant, called dark energy here. The scaling behavior of the mass density, the R–W scale, $a(t)$, the conformal time, and the Hubble parameter are shown in Table B.1.

Table B.1: Power law behavior of density, scale factor, conformal time and H.

	$\rho(a)$	$a(t)$	Conformal time τ	H
Matter	$1/a^3$	$t^{2/3}$	$t^{1/3}$	$(2/3)/t$
Radiation	$1/a^4$	$t^{1/2}$	$t^{1/2}$	$(1/2)/t$
Dark energy	constant	e^{Ht}	$\sim -1/(aH)$	constant

Explicit solutions for $\alpha(t)$ in the case of dominance of a single energy source:

$$\alpha_m(t) = [(3/2)\sqrt{\Omega_m}H_o t]^{2/3}$$
$$\alpha_\gamma(t) = [2\sqrt{\Omega_\gamma}H_o t]^{1/2} \tag{B.1}$$
$$\alpha_\Lambda(t) = \exp(\sqrt{\Omega_\Lambda}H_o t)$$

Appendix C

Symbol Table

Symbol	Definition
$\hbar$	Planck constant — reduced
$\alpha = e^2/\hbar c$	Electromagnetic coupling constant $\sim 1/137$
$\vec{p}$	Momentum vector
N_e	Number, Number of e-folds, $\ln(a)$
ℓ, S_ℓ	Lagrange density, action
ξ	Third slow roll parameter
$\langle \varphi \rangle$	Vacuum expectation value of Higgs field
$a(t)$	R–W scale factor
B	Binding energy
b	Impact parameter
$C_{\ell\ell}$	Projected angular CMB fluctuations
c	Speed of light
c_s	Speed of sound
ds	R–W metric
D, d	Physical, comoving distance
D_A, D_L	Angular, luminosity distance
De, Do	Distance at emission and reception
D_H	Present Hubble distance
e	Particle energy, eccentricity
f	Distribution function

(Continued)

(*Continued*)

Symbol	Definition
G	Newton constant
g^*	Number of degrees of freedom
G_F	Fermi effective weak coupling constant
$g_{\mu\nu}$	metric
h	Metric deviation — small GR. GR angular momentum
H	Hubble parameter, Ho present H
$K,\ k$	Physical, comoving wave vector
k_B	Boltzmann constant
L	Luminosity, angular momentum
L_{ls}	CMB last scattering probability
M, m	Mass, particle mass
Mp, Lp, tp	Planck mass, length, and time
n	Number density
$n_s,\ n_t$	Spectral index of the power — scalar, tensor Slow roll scalar, tensor parameters
Ne	Number of "e-folds"
p	Pressure, momentum
P	Power, pressure
$P(k)$	Fourier power in a fluctuation
P_h	Power in gravity wave
P_Φ	Power in potential
P_H	Power in inflation field
q	Deceleration parameter
Q	Quadrupole moment
r	Radial coordinate comoving radius, tensor to scalar power ratio
R	System radius
R, ζ	Ricci scalar and coupling to Higgs
R_s	Schwarzschild radius
$S,\ s$	Entropy, entropy density
T	Temperature
t, t_o	Time, present time

Symbol	Definition
t_c	Classical time for gravitational collapse
t_{dec}, a_{dec}	Time, scale factor for CMB decoupling
t_{eq}, a_{eq}	Time, scale factor for R/M equality
T_o	Present temperature
T_R	Pre/Re heat temperature
U	System energy
U_T, U_G	Thermal, gravitational system energy
u	Energy density
V	Scalar potential, volume
v_e, v_o	Velocity at emission and reception
X_n, X_e, X_A	Fraction of neutrons, ionized p, nuclei
z	Redshift parameter ($1/\alpha - 1$)
$\alpha(t)$	R–W scale, normalized to present, $a(t)/a_o$
α_H	Scale when mode exits the horizon
α_i	Coupling constant for SM force i
β	v/c
$\Gamma = 1/\lambda$	Reaction rate — inverse mean free path
Δ	Fractional changes in RGE for top or Higgs
δ_h	Fractional density fluctuation — tensor
δ_H	Fractional density fluctuation — scalar
$\delta\varphi$	Scalar field fluctuation
$\delta\phi$	Field fluctuation from the mean
ε, δ	Slow roll parameters
η	Baryon to photon ratio, entropy
θ	Polar angle
$\Theta = \delta T/T$	Chemical potential for photons
κ	Curvature parameter, opacity
Λ	Cosmological constant, wavelength
λ	Comoving wavelength, mean free path
λ_H	Quartic Higgs coupling constant
λ_J	Jeans length — supported by pressure
μ	Chemical potential, molecular weight
ρ	Energy density, mass density

(*Continued*)

Symbol	Definition
ρ_c	Present critical energy density
σ	Cross-section, Stefan–Boltzmann constant
τ	Conformal time, period
τ_z	Conformal time since emission, z
ϕ	Scalar field, Higgs field. Newtonian potential
Φ, Ψ	Temporal and spatial terms in the metric
Ω	Density scaled to critical density
ω	Equation of state parameter, circular frequency

Appendix D

Reference

There are many superb references for cosmology. In addition, web browsers allow the reader to search for specific topics with ease. For these reasons, only a single reference is given here. It is the site of the Particle Data Group. This group has historically focused on the properties of fundamental particles and detectors. Recently, however, given the great increase in the quantity and quality of cosmological data, the PDG has expanded and made a comprehensive coverage of both cosmological experiment and theory. The site is at https://pdg.lbl.gov/. The headings for the present cosmological review are shown in the following figure. In fact, the PDG site has comprehensive reviews of the standard model of both particle physics (SMPP) and cosmology (SMC). This is appropriate since a combination of both is needed to understand the current status of cosmology. In fact, this text attempts to connect the relevant aspects of both when they overlap.

In addition, a very comprehensive set of references accompanies each of the topics. Between the specific detailed web browser searches and the comprehensive overview of the experimental state of the art, no further searches need be made.

Appendix E

Acronyms

ACT	Atacama Telescope
BAO	baryon acoustic oscillations
BB	big bang
BK	Bicep Keck
CERN	European Center of Nuclear Research
CMB	cosmic microwave background
DE	dark energy
DESI	Dark Energy Survey Instrument
DM	dark matter
GR	general relativity
GUT	grand unified theory
HBB	hot big bang
IBB	inflationary big bang
IR	Infrared
LD	dark energy, (lambda) dominated epoch
LHC	Large Hadron Collider
LSS	large scale structure
MD	matter dominated epoch
NR	non-relativistic
NSE	nuclear statistical equilibrium
RD	radiation dominated epoch
RGE	renormalization group equations
RW	Robertson–Walker

(*Continued*)

(Continued)

SDSS	Sloan Digital Sky Survey
SMC	standard model of cosmology
SMPP	standard model of particle physics
SPT	South Polar Telescope
SR	special relativity
susy	supersymmetry
vev	vacuum expectation value
WIMP	weakly interaction massive particle
SHO	simple harmonic oscillator

Supplementary Materials

The supplementary materials include MATLAB files containing raw programming codes of more than 50 scripts presented in the book. Readers are encouraged to download the files and manipulate the codes/graphics to facilitate their comprehension of the accompanying discussion in the book.

Online access is automatically assigned if you purchase the ebook online via www.worldscientific.com. If you have purchased the print copy of this book or the ebook via other sales channels, please follow the instructions below to download the files:

1. Go to: https://www.worldscientific.com/r/14047-SUPP or scan the below QR code.

2. Register an account/login.
3. Download the files from: https://www.worldscientific.com/worldsci books/10.1142/14047#t=suppl. For subsequent access, simply log in with the same login details in order to access.

For enquiries, please email: sales@wspc.com.sg.

Index

www.ingramcontent.com/pod-product-compliance
Ingram Content Group UK Ltd.
Pitfield, Milton Keynes, MK11 3LW, UK
UKHW050654170325
456354UK00007B/445